Tropical Meteorology

Tropical Meteorology

Editor

Deb Nath

Tropical Meteorology

Edited by **Deb Nath**

Printed in 2017

ISBN: 978-1-68117-158-6

Library of Congress Control Number: 2015936571

© 2016 by
SCITUS Academics LLC,
616, Corporate Way, Suite 2, 4766,
Valley Cottage, NY 10989

www.scitusacademics.com

Contents

Preface vii

Chapter 1   Physics as Final Opportunity to Prevent Harms Related to
             Theatricalization of Meteorology 1
             Mbane Biouele César

Chapter 2   Relationship between Size of Cloud Ice and Lightning in the
             Tropics 19
             Deen Mani Lal, Sachin D. Ghude, Jagvir Singh, and Suresh Tiwari

Chapter 3   The West African Sahel: A Review of Recent Studies on the
             Rainfall Regime and Its Interannual Variability 41
             Sharon E. Nicholson

Chapter 4   Interannual and Intraseasonal Variability in Fine Mode Particles
             over Delhi: Influence of Meteorology 139
             S. Tiwari, D. S. Bisht, A. K. Srivastava, G. P. Shivashankara, and R.
             Kumar

Chapter 5   Satellite and Ground Measurements for Studying the Urban Heat
             Island Effect in Cyprus 165
             Diofantos G. Hadjimitsis, Adrianos Retalis, Silas Michaelides,
             Filippos Tymvios, Dimitrios Paronis, Kyriacos Themistocleous, and
             Athos Agapiou

Chapter 6   Initialization of Tropical Cyclones in Numerical Prediction
             Systems 195
             Eric A. Hendricks and Melinda S. Peng

Chapter 7   Effects of Urban Configuration on Human Thermal Conditions
             in a Typical Tropical African Coastal City 221
             Emmanuel Lubango Ndetto and Andreas Matzarakis

             Citations 255
             Index 259

Preface

Meteorology, climatology, atmospheric physics, and atmospheric chemistry are sub-disciplines of the atmospheric sciences. Meteorology and hydrology compose the interdisciplinary field of hydrometeorology. Interactions between Earth's atmosphere and the oceans are part of coupled ocean-atmosphere studies. Meteorology has application in many diverse fields such as the military, energy production, transport, agriculture and construction. Fundamental science of the tropical atmosphere and synthesizes the tremendous increase in our knowledge of tropical meteorology during the past two decades. In that same period, great advances also occurred in learning technologies; allowing students to learn through interactivity and access to real data.

Editor

Physics as Final Opportunity to Prevent Harms Related to Theatricalization of Meteorology

Mbane Biouele César

Laboratory of Earth's Atmosphere Physics, Department of Physics, University of Yaoundé I, Yaoundé, Cameroun

ABSTRACT

Using Mbane Biouele formula derived in 2009 on the troposphere thermoelastic properties leads to thermal and kinematic profiles of major atmospheric disturbances which clearly indicate that these terrible events for men should not be viewed with fatalism. This unexpected truth is unfortunately always obscured by media outlets of brilliant TV presenters or famous workshops panelists that focus attention on the excessively sensational meteorology (unfortunately folk and pernicious) instead of worrying about the seriousness that should characterize all interventions on the climate study or prediction. Good weather conditions, it is undeniable, facilitate an excellent running of almost all human activities like sports, transport, agricultural activities, celebrations of events, etc.... Far more serious, the advent of supercomputers and satellites could, if their valuable information is used solely for the theatricalization of weather events, trigger the decline of the scientific discipline of great public utility that is meteorology. Indeed, many meteorologists acquire very big head when they succeed in acquiring advanced equipment. Without prejudging what meteorology will become in the future, we hope that the work done in this article will remind each researchers that much remains to be done to promote climate studies. We remind quite emphatically that both hurricanes and cyclones have their weak-points (or talon d'Achilles in French) and thus, researchers should begin to think about "how to neutralize atmospheric disturbances that have both a large and a strong destructive power".

INTRODUCTION

The advent of supercomputers and satellites could, if their valuable information is used solely for the theatricalization of weather phenomena, trigger the decline of meteorology. Indeed, many meteorologists become proud to excess (similar to cinematograph films stars) from the time they work in structures with worldwide reputation. These celebrities of a different kind refuse any contradiction (including the most relevant) and unfortunately convey erroneous and inconsistent information during their numerous interventions in the workshops. Fortunately, physics appears today as final opportunity to prevent harms related to excessive theatricalization of meteorology. Fortuitously, the application of Mbane Biouele formula derived 2009 on troposphere thermoelastic properties [1-5] and its related corollaries lead to appropriate and unique profiles of major weather disturbances which show without any dough for us that both cyclones and tornadoes have their weak-point or talon d'Achilles and should not be viewed with fatalism. We hope that this unexpected reality will remind each researcher that much remains to be done to promote our knowledge of the atmosphere disturbances that have both a large and a strong destructive power.

MBANE BIOUELE FORMULA ON THE TROPOSPHERE THERMOELASTIC PROPERTIES

Atmosphere Dynamics Concept of Air Particle

Atmosphere Dynamics use a set of assumptions (previously uncontested) to define the particle or parcel of air [6-8]. Especially:

- Few exchanges on molecular scale: it is easy to follow quantity of air which preserves certain properties.
- Quasi-static equilibrium: at any moment there is dynamic balance, i.e., the particle has the same pressure as its environment ($P = P_{ext}$).
- No thermal balance: heat transfers by conduction are very slow and neglected. One can have $T \neq T_{ext}$.
- The horizontal sizes of the air particle can go from a few cm to 100 km according to the applications.

Mbane Biouele Formula on the Troposphere Thermoelastic Properties

Taking into account the fact that the atmosphere is mainly composed of dry air and water vapor, the Dalton's law connects the pressure (P) with the partial pressure of dry air (P_a) and saturated water vapor (e_w)

$$P = P_a + e_w$$

(1)

In deriving (P) with respect to the temperature, one has

$$\frac{dP}{dT} = \left(\frac{\partial P}{\partial T}\right)_V + \left(\frac{\partial P}{\partial V}\right)_T \left(\frac{dV}{dT}\right)$$

(2)

According to the Quasi-static equilibrium (or dynamic balance) the pressure of the parcel of air must be the same as that of the ambient air, including during sudden local changes in phases by water contained in this parcel. Reality that all meteorologists restore through the following famous sentence: the phases change takes place at constant pressure. Thereby write Equation (3),

$$dP = 0$$

(3)

Equations (2) and (3) lead to the derivative of V compared to T

$$\frac{dV}{dT} = -\frac{\left(\dfrac{\partial P}{\partial T}\right)_V}{\left(\dfrac{\partial P}{\partial V}\right)_T}$$

(4)

Introducing the coefficient of thermal expansion of moist air at constant temperature

$$\chi = -\frac{1}{P}\left(\frac{\partial P}{\partial V}\right)_T$$

(5)

Then the Fundamental Relationship of Atmosphere Dynamic Balance:

$$\frac{dV}{dT} = \frac{1}{\chi} \bullet \frac{1}{P}\left(\frac{\partial P}{\partial T}\right)_V$$

(6)

One can also write equation of Atmosphere Dynamic Balance in terms of partial pressures

$$\frac{dV}{dT} = \frac{1}{\chi} \bullet \frac{1}{P}\left[\left(\frac{\partial P_a}{\partial T}\right)_V + \left(\frac{\partial e_w}{\partial T}\right)_V\right]$$

(7)

$\left(\frac{\partial P_a}{\partial T}\right)_v$ is negligible compared to $\left(\frac{\partial e_w}{\partial T}\right)_v$

Thereby write formula (8) named Mbane Biouele formula on troposphere thermoelastic properties

$$\frac{dV}{dT} \cong \frac{1}{\chi} \bullet \frac{1}{P}\left[\left(\frac{\partial e_w}{\partial T}\right)_V\right]$$

(8)

Formula (8) Leads To A Meteorological Very Useful Statement:

At any moment and throughout the atmosphere, one can use formula (8) and Clausius-Clapeyron slope of the equilibrium curves in the eT-diagram (Figure 1) to predict in which direction the air parcel will move (up or down) if its temperature increases or decreases. Table 1 or Figure 2provides an overview of possible situations throughout the Troposphere.

EVIDENCE OF EFFECTIVENESS OF MBANE BIOUELE FORMULA (2009)

The two equal level surfaces of water vapor and temperature rating respectively at 6.11 mb and 0.0098°C

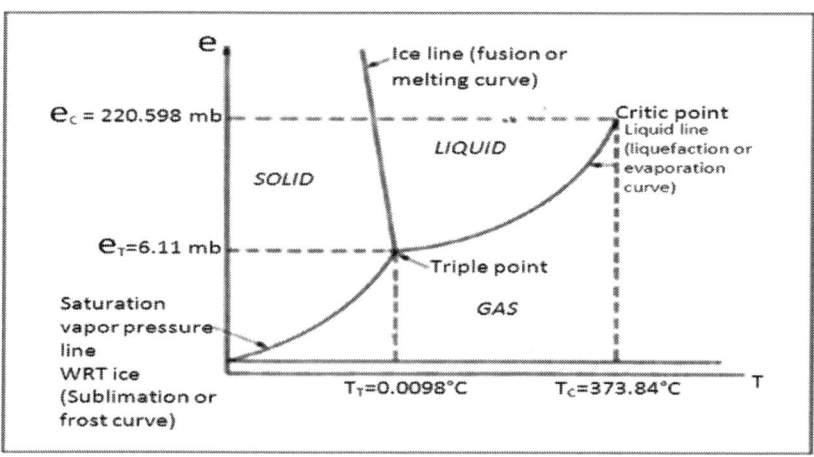

Figure 1: Saturation curves for water substance onto the eT-plane (e_{wT} and T_T are triple-point coordinates): e_{wT} = 6.11 mb; T_T = 0.0098°C.

Table 1: Changes in volume of the moist air particle depending on temperature within a specific range of temperature and humidity

Range of temperature coupled with range of humidity	T < 0.0098°C ew < 6.11 mb	T < 0.0098°C ew > 6.11 mb	T > 0.0098°C ew > 6.11 mb
$\left(\dfrac{\partial_{e_w}}{\partial T}\right)_v$	+	-	+
$\dfrac{dv}{dt}$	+	-	+

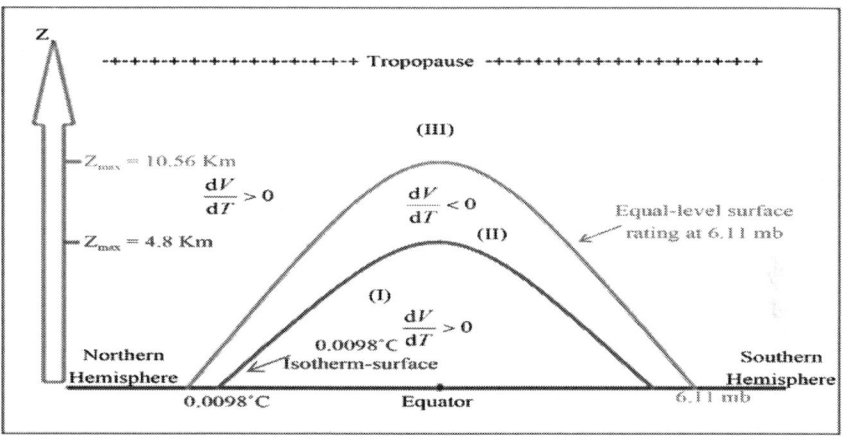

Figure 2: Troposphere specific regions depending on the manner in which V changes with T (V and T are respectively volume and temperature of an air parcel): If (dV/dT) > 0 the particle swells when its temperature increases (so it becomes lighter), if (dV/dT) < 0 the particle shrinks when its temperature increases (so it becomes less light). Z_{max} are statistical value of observed maximum elevation of equal level surfaces of temperature and water vapor rating respectively at 0.0098°C and 6.11 × 10⁻³ bars.

(Figure 2) separate clearly parts of the troposphere where ideal gas assumption can be applied without too distort reality to parts of the troposphere where this assumption is banned: hence appropriate and unique plots of thunderclouds profiles of wind and electric charges and earth's atmosphere tricellular circulation.

Thunderclouds Winds' Profiles and Related Electrical Charges

All natural meteorological phenomena included Hurricanes and Cyclones can be traced to the manner in which the energy from the sun is received over different parts of the earth. Since the troposphere is a medium in which mass motions are easily started, convection is found to be one of the chief ways in which heat is transferred there. This transfer may be accomplishing either by vertical or by horizontal motions. According to our results: warmer disturbances that occur in lower-troposphere are dissipated by a typical mass motion usually called Hurricanes (or tornadoes) while cooler disturbances that occur in mid-troposphere are dissipated by another typical mass motion called Cyclones. Knowing that Coriolis force act to west on updrafts, everyone can now understand why Hurricane and Cyclone move preferentially from East to West due to the localization in updrafts of their heat sources. Cyclones' heat source is made of huge and cooler fogs (those observed temperatures are less than −45°C) which can travel even increase (over hot oceans) in the troposphere while hurricanes' heat source is fixed on the Ground: that's why Cyclone lives and travels longer than Hurricane.

a) Schematic Representations of Hurricanes and Related Cloud or Electrical Systems

Hurricanes appear (Figure 3(a)) as very high towers (from 0 to about 9 Km) consisting of three floors: warm updrafts occupying the first and third while warm downdrafts occupies the second floor. According to ground based observations, over-land hurricanes (or tornadoes) trigger thunderclouds whose base is thin compared to the peak which is very broad. Lower troposphere updrafts of Figure 3(a) can, in the same place, simultaneously trigger (due to wind-stress and adiabatic expansion) electrical positive charges and dark-cloud (Stratus) as suggested in Figure 3b)(. The broadest peaks of the related clouds indicate the presence of the second floor warm downdrafts that prevent the progression of the first floor warm updrafts. Figure 3(a) is completely different to theatrical Figure 3(c) which contains many contradictions. e.g.: 1) Tornado or Cyclone eyes are known as a low pressure: the diagram says otherwise placing the eye at the center of divergence; 2) the veil of thick and very high white cloud (named Cumulonimbus) that envelops the black cloud formation is present nowhere in the earth's atmosphere (including on pictures taken on Tornadoes); 3) downdrafts generally do not create the clouds and cannot trigger tornadoes dark clouds as shown in the diagram.

b) Schematic Representations of Cyclones and Related Cloud or Electrical Systems

Cyclones appear (Figure 4(a)) as very high towers (from 0 to about 14 Km) consisting of three floors: cooler downdrafts occupying 1st and 3rd while cooler updrafts occupy the 2nd floor. There is good agreement between aircraft-based observations and related cyclones second floor updrafts convective clouds whose base has to be located above $0.0098°C$ isotherm surface. Mid troposphere downdrafts of Figure 4(a) can, in the same place, simultaneously trigger (due to wind-stress and rapid-cooling of troposphere lower layers below their dew point temperature) electrical positive charges and darkcloud (Stratus) as suggested in Figure 4(b). View from the ground surface, the three cloud systems of Figure 4(b) appear to the observer as a single very high vertical extension cloud named Cumulonimbus by meteorologist. The same clouds system, seen from an airplane flying at 7000 meters, is as shown in Figure 4(b). i.e.: Low-altitudes stratus above which is placed a single cumulus whose base is broader than its top.

c) Schematic Representations of Hurricane and Cyclone Horizontal Winds

Observed pressure near the eyes of Cyclones (or hurricane) is very low and concentrates a rapid decrease in a short distance so that the momentum of particles of air, the frictional force and the tidal force are (from surface of the earth to tropopause) negligible compared to the Coriolis and pressure-gradient forces. When pressure gradient and Coriolis forces are the only two factors acting, geostrophic winds (rotative in the Northern hemisphere and contra-rotative in the Southern hemisphere) immediately take place (Figures 5) within deep and passive convections. The impact of hurricanes footprint (less than a dozen kilometers in diameter) is much lower than that of cyclones (several tens of kilometers in diameter).

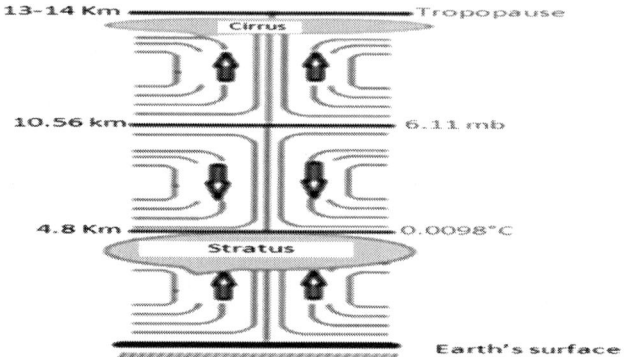

Vertical profile of Hurricane and related clouds system
(Mean statistical values are in dark color while
Thermodynamic values are in blue color)

(a)

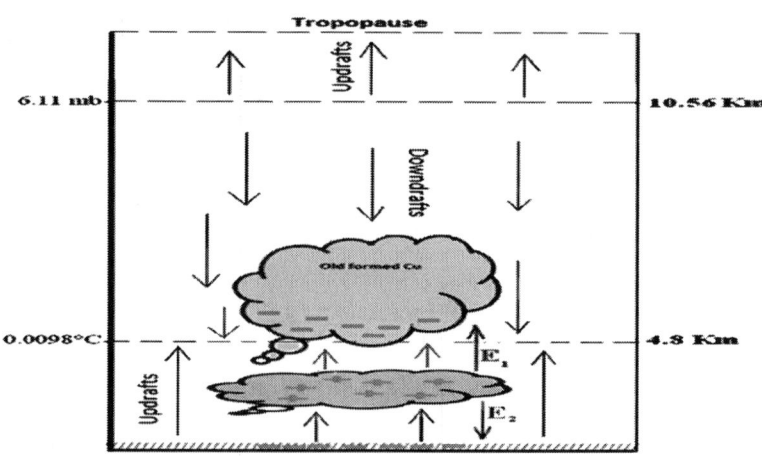

(b)

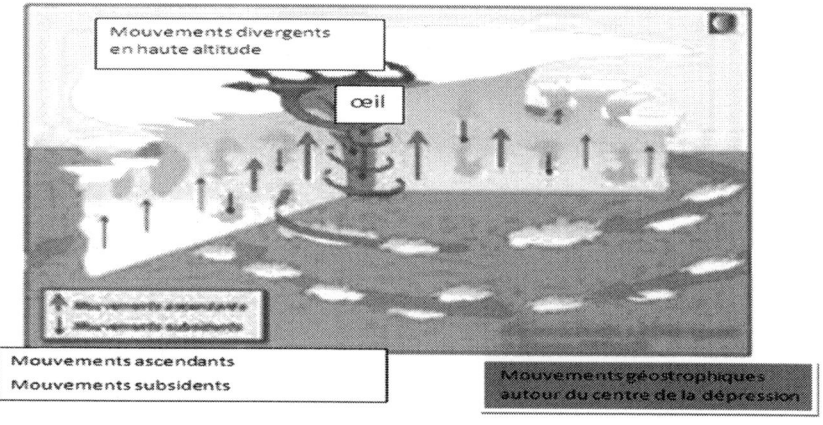

(c)

Figure 3: Schematic representations of Hurricanes and related cloud or electrical systems. Hurricanes' new-formed clouds are gray while old-formed clouds are blue.

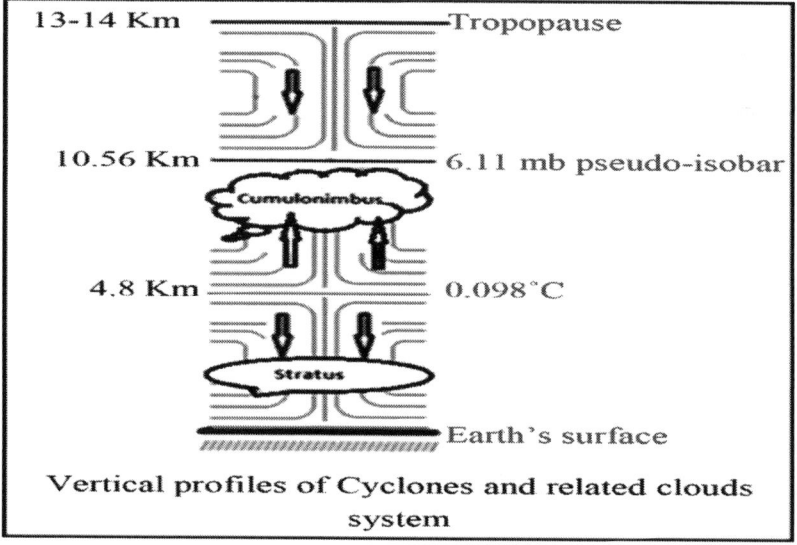

(a)

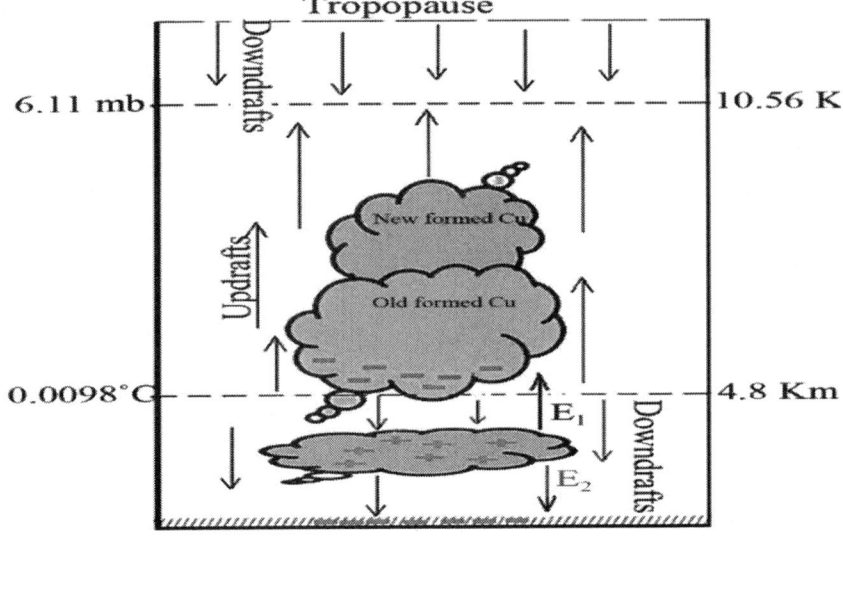

(b)

Figure 4: Schematic representations of Cyclones and related cloud or electrical systems. Cyclones' new-formed clouds are gray while old-formed clouds
are blue.

d) Thermodynamic or Chemical Constraints on Rain

Clouds are physical systems consisting of water particles. Even if the
volume of a water particle is large; as long as its density $r_w(z)$ remains
less than that of the surrounding air $r_A(z)$, the particle remain suspended
in the air. For causing its fall: the water particle must be contaminated
with a soluble chemical substance having density greater than r_A (at z =
0). Natural Contaminations are performed with carbon dioxide; sulfur
emitted by volcanoes or sprays emitted by the oceans, etc. Acid rains
reflect this thermodynamic constraint. Clouds separate themselves with
favorably contaminated water particles and that give rise to rainfall or
snowfall.

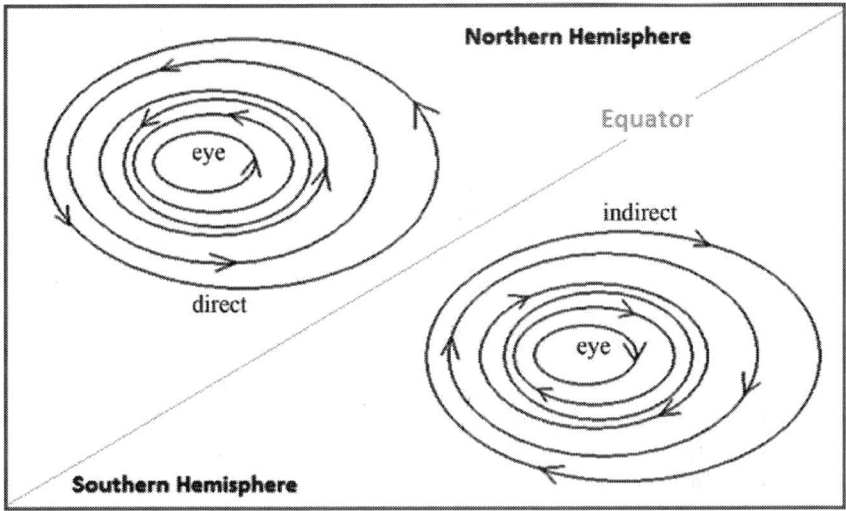

Figure 5: Schematic representations of Cyclones and Tornadoes geostrophic balance winds.

Relationship between Cloud Formation's Additional Green House and Tornadoes

Considering the molecular scale, our model (Figure 6) based on Mbane Biouele formula (2009) suggests blocking of hot updrafts by hot down drafts which means installation of an additional greenhouse that triggers the superposition of cloud formation latent heat with earth's surface radiate heat R_T $(R_T = \varepsilon_s \sigma T_s^4)$. This is consistent with based observations and explains high surface temperatures that accompany the formation of clouds in the sunny sky. Furthermore, Tornadoes look like little-bombs triggered by cloud's additional greenhouse. Indeed, Rayleigh and Reynolds numbers of passive convection motions depend mainly on air parcel temperature and humidity. Within additional greenhouse, temperature and water vapor increase exponentially and finally (as atmosphere is a dissipative system) trigger violent adiabatic expansion or tornadoes (as Rayleigh and Reynolds numbers become higher and generate turbulent motions). Thermodynamic processes that govern the formation of tornadoes or hurricane (Figures 3(a) and (b)) are well known now.

Appropriate and Unique Representation of Earth's Atmosphere General Circulation

According to our previous study [9], the presence of water substance in all three states in the earth's atmosphere gives to troposphere the exclusivity of a general circulation consists of three groups of passive convective cells (Hadley, Ferrel, and Polar) on either side of the ITCZ (Inter Tropical Convergence Zone). These cells (H, F, and P) take place within areas bounded by both tropopause and the two equal level surfaces of water vapor and temperature rating respectively at 6.11 mb and 0.0098°C as suggested in Figure 7.

CONCLUSIONS AND COMMENTS

All natural meteorological events including hurricanes and cyclones can be traced to the manner in which the energy from the sun is received over different parts of the earth. Since the troposphere is a medium in which mass motions are easily started, convection is found to be one of the chief ways in which heat is transferred there. This transfer may be accomplished either by vertical or by horizontal motions. According to our results: warmer disturbances that occur in lower-troposphere are dissipated by a typical mass motion usually called Hurricanes (or tornadoes) while cooler disturbances that occur in mid-troposphere are dissipated by another typical mass motion called Cyclones. Thermal and Kinematic profiles of atmosphere disturbances that have both a large and a strong destructive power clearly indicate that these terrible events for men should not be viewed with fatalism. Viewing the urgency to perform our knowledge of tornadoes or cyclones and erasing our great fear of these events, theatricalization of meteorology becomes something unacceptable. Our research field (meteorology in this case) can be popularized while remaining serious about what is presented to those who consume it. We hope that one day the weak-points of hurricanes and tornadoes will be exploited to neutralize them. e.g., although costly, procedures to extinguish the huge forest fires are developed and used successfully. It is possible, we reflect deeply on this issue, that the neutralization of tornadoes requires less effort than those made to forest fires.

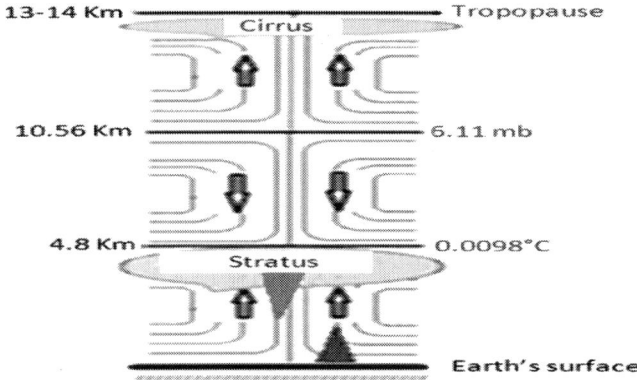

Mbane Biouele formula (2009) suggests blocking of hot updrafts by hot down drafts: then installation of an additional greenhouse which triggers the superposition of cloud formation latent heat (red color) with earth's radiating heat $e_s sT^4$ (brown color). That explains high surface temperatures that accompany the formation of clouds in the sunny sky (or in extreme cases: the formation of tornadoes)

Figure 6: Clouds formation additional green house.

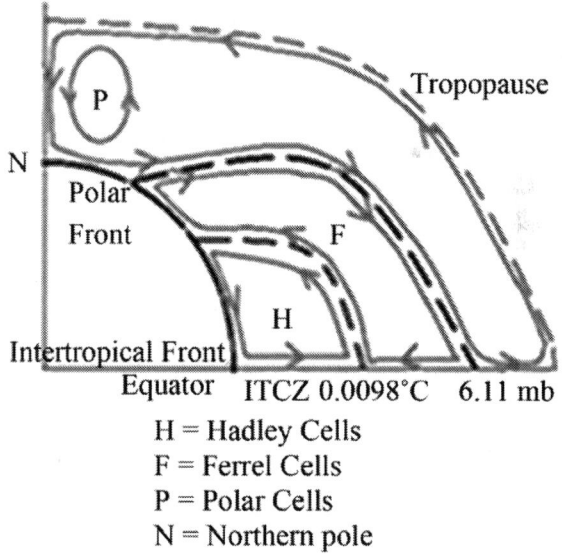

H = Hadley Cells
F = Ferrel Cells
P = Polar Cells
N = Northern pole
Streamlines of Earth's Atmosphere General Circulation

Figure 7: Schematic representation of the General Circulation, derived from Mbane Biouele formula (2009) and related corollaries.

ACKNOWLEDGEMENTS

Author of this paper gratefully acknowledge the following University Institutions and Professors: 1) The University of California in Los Angeles (USA) who kindly provided a geospatial measurement station called "AMBER MAGNETOMETER" to the University of Yaoundé I (Cameroon); The Air Force Research Laboratory (Boston University), which has equipped the University of Yaoundé I with two geospatial measurement stations called "SCINDA SENSORS" and The University of Yaoundé I which offered modern facilities with Internet Connection, to accommodate both stations. 2) Professors: Mark B. Moldwin (Atmospheric-Oceanic and space Sciences Research Laboratory, University of Michigan (USA)) and Endawoke Yizengaw (Institute for Scientific Research (USA)), for their very kind collaboration.

REFERENCES

1.   B. Couanang Siebatcheu, C. Mbane Biouele and J. S. Eyebe Fouda, "Atmosphere Dynamic Balance Model (ADBModel) and Related Troposphere General Circulations' Cells behind the Formation of Tropical Monsoons," Scholars Research Library—Archives of Physics Research, Vol. 3, No. 2, 2012, pp. 93-100.

2.   C. Mbane Biouele, "Hurricanes and Cyclones Kinematics and Thermodynamics Based on Clausius-Clapeyron Relation Derived in 1832," International Journal of Physical Sciences, Vol. 8, No. 23, 2013, pp. 1284-1290.

3.   C. Mbane Biouele, E. Yizengaw, M. B. Moldwin and G. Cautenet, "Impacts of Thermoelastic Properties of Saturated Water Vapor on Tropical Depressions Thermodynamics and Dynamics," Scholars Research Library—Archives of Physics Research, Vol. 2, No. 4, 2011, pp. 24-33.

4.   C. Mbane Biouele, "Vertical Profiles of Winds and Electric Fields Triggered by Tropical Storms-Under the Hydrodynamic Concept of Air Particle", International Journal of Physical Sciences, Vol. 4, No. 4, 2009, pp. 242- 246.

5.   C. Mbane Biouele, "Physics of Atmosphere Dynamic or Electric Balance Processes Such as Thunderclouds and Related Lighting Flashes," Geosciences, Vol. 2, No. 1, 2012, pp. 6-10.

6. H. R. Byers, "General Meteorology," McGraw-Hill Book Company, INC., 1959, 540 p.

7. G. K. Batchelor, "An Introduction to Fluid Dynamics," Cambridge University Press, 1967, 496 p.

8. C. A. Riegel, "Fundamentals of Atmospheric Dynamics and Thermodynamics," World Scientific Publishing Co. Pte. Ltd., 1992, 512 p.

9. C. Mbane Biouele, "Application of Clausius-Clapeyron Relation (1832) and Carnot Principle (1824) to Earth's Atmosphere Tricellular Circulation," Atmospheric and Climate Sciences, Vol. 4, No. 1, 2014, pp. 1-6.

Chapter 2

Relationship between Size of Cloud Ice and Lightning in the Tropics

Deen Mani Lal[1], Sachin D. Ghude[1], Jagvir Singh[2], and
Suresh Tiwari[1]

[1]Indian Institute of Tropical Meteorology, Pune 411008, India
[2]Ministry of Earth Sciences, Lodhi Road, New Delhi 110060, India

ABSTRACT

The association of lightning flashes with mean cloud ice size over continental and oceanic region in the tropical areas has been analyzed using the observations from various satellite platforms (MODIS, TRMM, and LIS) for the period 2000–2011. We found that frequency of lightning in general is higher over the continental region compared to oceanic region, whereas larger size of cloud ice is observed over the oceanic regions compared to the continental regions. Relationship between lighting and cloud ice size shows similar features over both continental and oceanic regions. For the first time, we show that total lighting increases with increase in the cloud ice size; attends maximum at certain cloud ice size and then decreases with increase in cloud ice size. Maximum lightning occurred for the mean cloud ice size of around 23–25 μm over the continental region and mean cloud ice size of around 24–28 μm over the oceanic region. Based on our observation we argue that the relation between lightning and mean cloud ice size follow the curve linear pattern, and not linear.

INTRODUCTION

Generation of lightning in atmosphere is still a matter of debate. It is a commonly established fact that ice is a key element to generate and separate the positive and negative charges inside the cloud which assists formation of lightning in the atmosphere [1]. During the occurrence of deep convection, water vapors are uplifted and condensed to form the deferent sizes of no inductive hydrometers (ice crystal, hail, drops, etc.); afterwards they are evaporated/sublimated and dispersed zonally and meridionally in the upper troposphere [2]. During upward motion, hydrometers collide with each other generating the charge on ice crystals, graupel, and liquid water [3, 4]. Some of the earlier studies [3, 5] have shown that approximately 5×10^{-4} e.s.u. (0.17 pC) charges are transferred per collision between the crystals and graupels of radius approximately 50μ m. This activity mostly takes place at the height where the temperature is colder than -10°C [6]. Similarly, in other studies based on field measurement the charge density around 1 to 10 C/km³ at the levels between the isotherms of −10 to -25°C have been reported [7]. Carey and Rutledge [8], Petersen et al. [9, 10], and Dye et al. [11] have found that strong updraft and production of significant lightning occurs at the height where the temperature is in between 0°C to -40°C. It is also believed that most of the noninductive charges inside the thunderstorm (due to rebounding collisions between graupel and ice crystals in the presence of super-cooled liquid water) are generated in this temperature region [4, 12–18]. The separation of the charge among the particles inside a cloud depends on the relative motion of hydrometers, whereas the rate of charge transfer and polarity depends on the size of particle, temperature and the liquid water content [19–21]. In addition, charges are also generated on crystal during condensation, evaporation, and sublimation/melting of ice. The rapid growth of electrification has been reported when cloud particles are frozen and form the ice during updraft [22, 23]. Laboratory experiment [3] showed that graupel pellets gain much more charges if graupels are growing by riming and collide with ice. Some laboratory experiment also showed that during freezing of distilled water very small negative charges are generated, whereas during melting much larger negative charges are generated [24, 25], but this effect could not be observed when water is contaminated.

Aerosol also affects the cloud ice concentration and its size by reducing the mean droplet size, which enhances the ice concentration in the region where temperature is less than zero [26]. Takahashi [27] found that increasing tendency of lightning flashes is positively correlated with increasing concentration of cloud ice as well as its size. Sherwood et al. [28] reported that occurrence of maximum lightning is associated with small size of cloud ice. In another study, decreasing size of cloud ice with increasing the aerosol concentration has also been reported [29].

In recent study over central India [30] a positive correlation has been found between ice concentration and lightning during premonsoon and monsoon seasons, whereas Deierling et al. [1] reported significant correlation between both precipitation and nonprecipitation ice mass with total lightning over Northern Alabama and Colorado/Kansas. Similar relation between lightning and cloud ice masses has also been reported in other field observations [1, 10, 17, 31–34]. The combined effect of aerosol with thermodynamic effect over India [35] and threefold enhancement of cloud-to-ground lightning flash density over Houston, Texas [36], raises the issue of pollution or heat island effect as a cause. As ice is a form of frozen cloud drops above the freezing level during deep convection, some results reported positive relation of lightning with strong updraft [11, 37–39].

Sizes of cloud ice represent the meteorological condition, aerosol effect, atmospheric dynamics, and are closely related to the cloud electrification and lightning discharge. It is still not clear whether small or large ice sizes increase the lightning flashes. Sherwood et al. [28] reported that small ice generate more lightning, whereas less cloud electrification in small cloud ice area is also reported [18, 40]. In this study we have presented the relationship between size of cloud ice and lightning flashes on global scale (over tropical regions). Lightning flashes and effective radius of cloud ice are considered over both continental as well as oceanic region.

DATA

In this study monthly mean cloud effective radius of ice phase (cloud particle size) (QA-W) from Moderate Resolution Imaging Spectroradiometer (MODIS) Level-3, cloud ice concentration from 3A12 version 6, and area averaged lightning from Lightning Imaging Sensor (LIS) on board of Tropical Rainfall Measurement Mission's (TRMM) satellite for the period of 2000–2011 data sets have been used for analysis. MODIS Level-3 was first launched on 18 December, 1999, on board the Terra platform and subsequently on 4 May, 2002, on board the Aqua platform, which is uniquely designed (high spatial resolution, wide spectral range, and near daily global coverage) to observe and monitor cloud effective radius and other Earth changes. We have used MODIS $1° \times 1°$ gridded level-3 monthly averaged cloud particle size from Terra platform (2000–2011) over continental region shown in Figure 1(d) as L1 [lat. (−35)–(−22), long. (−67)–(−47)], L2 [lat. (−12)–(−8), long. 9–29], and L3 [lat. 24–34, long. 68–83] in our analysis. Similarly, we used MODIS $1° \times 1°$ gridded level-3 monthly average cloud particle size data set from Aqua (2002–2011) over the oceanic region shown in Figure1(d) as O1 [lat. (−32)–(−20), long. (−146)–(−126)], O2 [lat. 12–22, long. 49–73], and O3 [lat. 0–18, long. 124–149]. We have also used MODIS, Terra platform data set, for the period of 2000-2001 for oceanic region. The agreement between MODIS monthly average cloud particle size data product from Terra and Aqua is observed to be 90% (R=0.95) suggesting that both data sets are quite consistent. TRMM Microwave Instrument (TMI) profiling gives global vertical hydrometer profiles and surface rainfall mean on $.05° \times 0.5°$ grid resolution. This data set is available at TRMM Online Visualization and Analysis System (TOVAS) web-based interface (http://gdata1.sci.gsfc.nasa.gov/daac-bin/G3/gui.cgi?instanceid=TRMM_Monthly). We have used area average monthly vertical profile of cloud ice concentration retrieved by 3A12 algorithm over the continental region (L1, L2, and L3) and oceanic region (O1, O2, and O3). We have also used lightning data from LIS which is a science instrument on board the TRMM observatory launched on 28 November, 1997. The detection efficiency of LIS is more than 80% in both daytime and nighttime with resolution (4 to 7 km) over a large region (600×600km) of the Earth's surface for total lightning (i.e., intracloud + cloud-to-ground) [41].

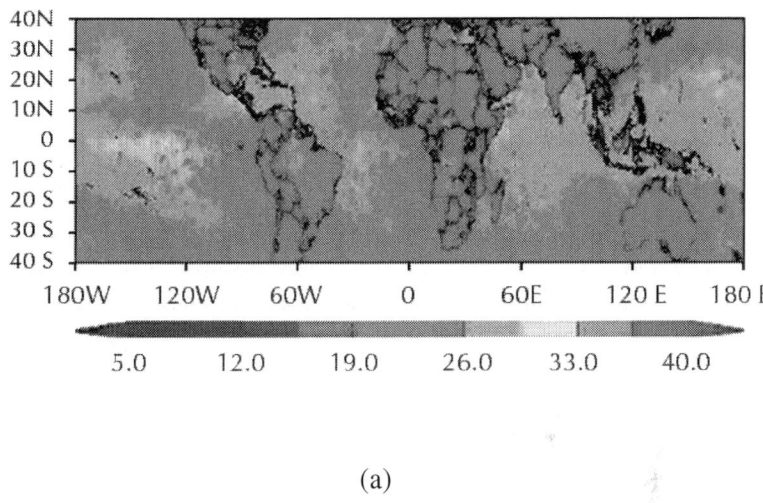

(a)

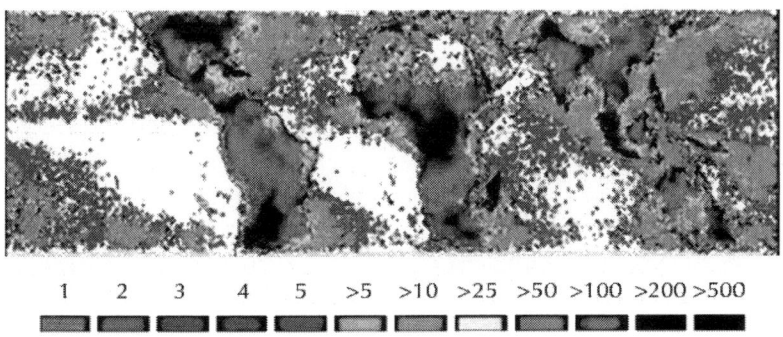

(b)

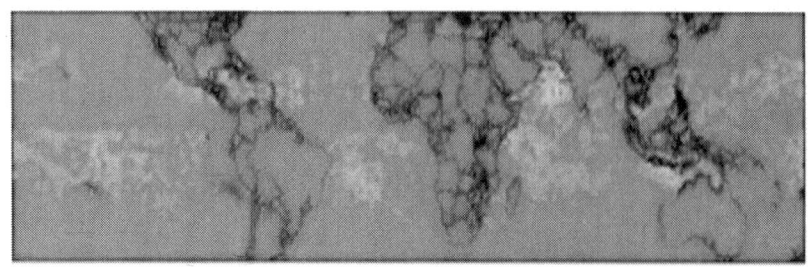

(c)

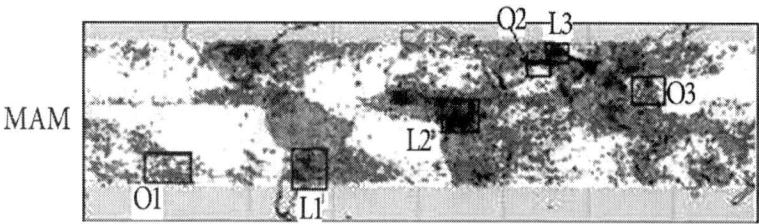

MAM

(d)

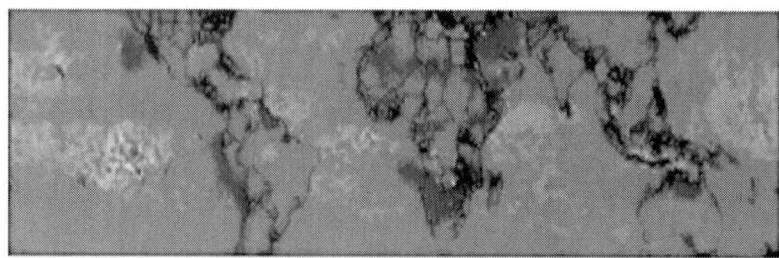

(e)

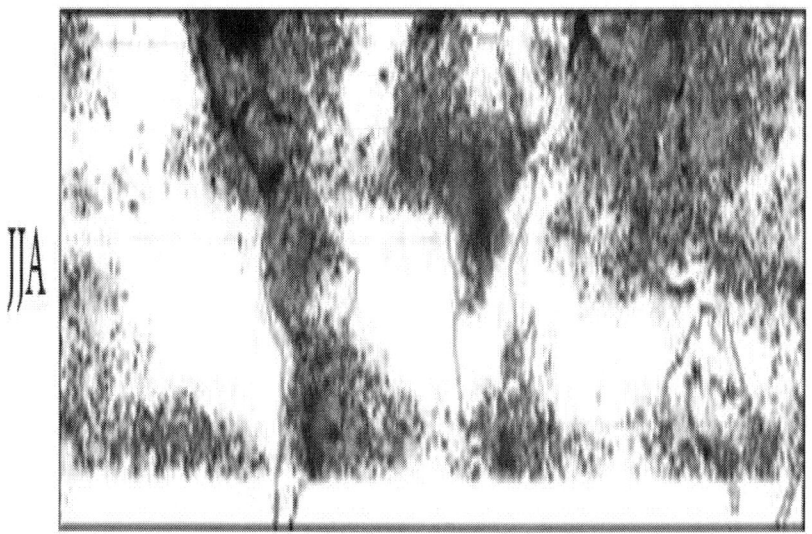

(f)

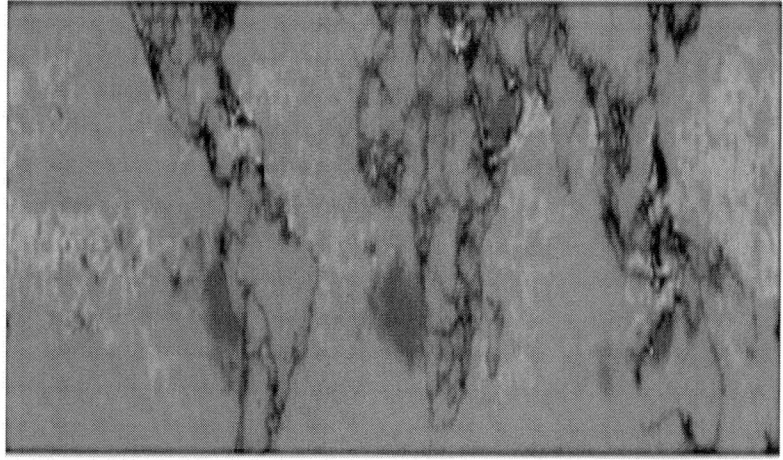

(g)

SON

(h)

Figure 1: Spatial distribution of (a) size of cloud ice and (b) lightning for the annual average and four seasons during 2005.

Selection of Study Area

Figures 1(a) and 1(b) show the example of spatial distribution of size of cloud ice and lightning, respectively, for the year 2005. Year 2005 is neither drought year nor very heavy rainfall year. It is considered as normal rainfall year [42]. In order to study the association between size of cloud ice and lightning, we have selected three regions over the continental region as L1, L2, and L3 and over the ocean as O1, O2, and O3 shown in Figure 1(d). We have used 12 years (2000–2011) of monthly cloud ice size and lightning data over the study area (L1, L2, L3 and O1, O2, O3) for analysis.

RESULTS

Spatial Distribution of Cloud Ice Size and Lightning

Figure 1 shows the spatial and seasonal pattern of lightning and cloud effective radius (cloud particle/ice size) over the tropical regions for the year 2005. A clear spatial change in lightning and cloud ice can be seen in Figure 1. It is interesting to note from Figure 1 that frequency of lightning in general is higher over the continental region compared to oceanic region. The annual average lightning flashes over the areas L1, L2, and L3 are generally greater than 500 flashes/km^2/month, whereas over the areas O1, O2, and O3 are less than 25 flashes/km^2/months. On the other hand, larger size of cloud ice is observed over oceanic region compared to continental regions. The average cloud ice size is greater than 30 μm (some places more than 40 μ m) over the oceanic region (O1, O2, and O3) and less than 25 μ m over the selected areas (L1, L2, and L3) on continental region. During spring months (March, April, and May; MAM), intense lightning can be seen (Figure 1(d)) over the continental regions such as Uruguay and surrounding regions (East part of Argentina and south Brazil), central part of United State, Colombia, Central African Republican and surrounding region, Democratic Republican of Congo (DRC), eastern part of South Africa, India (Indo Gangetic plain and in some other parts), South-East part of China, Thailand, and Indonesia. Lightning frequency greater than 150 flashes/km^2/month (Figure 1(d)) and the average effective cloud ice diameters between 22 and 25 μ m (Figure1(c)) have been observed over these regions. The lightning frequency over Mexico, Guatemala, Nicaragua, Angola, Namibia, and the entire part of Brazil during spring months has been observed to be less than 25 flashes/km^2/month (effective cloud ice diameter 28–31 μ m). In comparison with continental regions (discussed above) the low lightning frequency (<5 flashes/km^2/month) and large ice particle size (31–40 μ m) has been observed over the oceanic region. In Figure 1, the similar features are also evident during summer monsoon (June, July, and August; JJS), winter (December, January, and February; DJF), and fall months (September, October, and November; SON).

Relationship between Cloud Ice Size and Lightning

In order study the relationship between cloud ice size and lightning in detail we have analyzed the monthly mean cloud effective radius and total lightning (for the period 2000–2011) averaged over the continental (L1, L2, and L3) and oceanic (O1, O2, and O3) areas shown in Figure 1(d). Months corresponding to the cloud ice size between 19 and 34 μm are grouped in the bin size of 1μm (we have considered all the months (12×12=144 months) for frequency count). We have noticed that there were hardly any months in which monthly mean cloud ice size was less that 19μm or greater than 34 m during the study period. Lightning corresponding to each bin (of 1 μm) is then added to obtain total lightning for every 1μm bin between 19 and 34 μm. Figure 2 shows relationship between cloud ice size and lighting over the three different continental (Figure 2(a)) and oceanic regions (Figure 2(b)). It can be seen that relationship between lighting and cloud ice size shows similar pattern over both continental and oceanic regions. Maximum lightning occurred for the mean ice cloud sizes of 24, 25, and 23 m over the continental regions L1, L2, and L3, respectively. Similarly, over the oceanic regions O1, O2, and O3 maximum lightning occurred for the slightly greater mean cloud ice size of 26, 24, and 28μm, respectively. It is interesting to note from Figure 2 that total lighting increases with increase in the cloud ice size, attains maximum at certain cloud ice size, and after that starts decreasing with increasing cloud ice size.

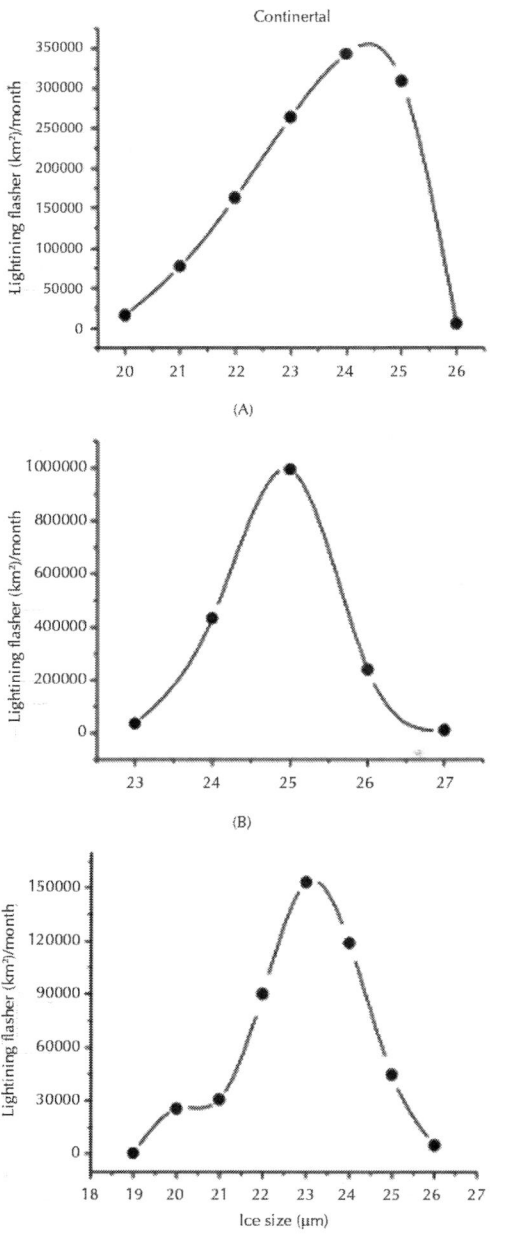

(a)

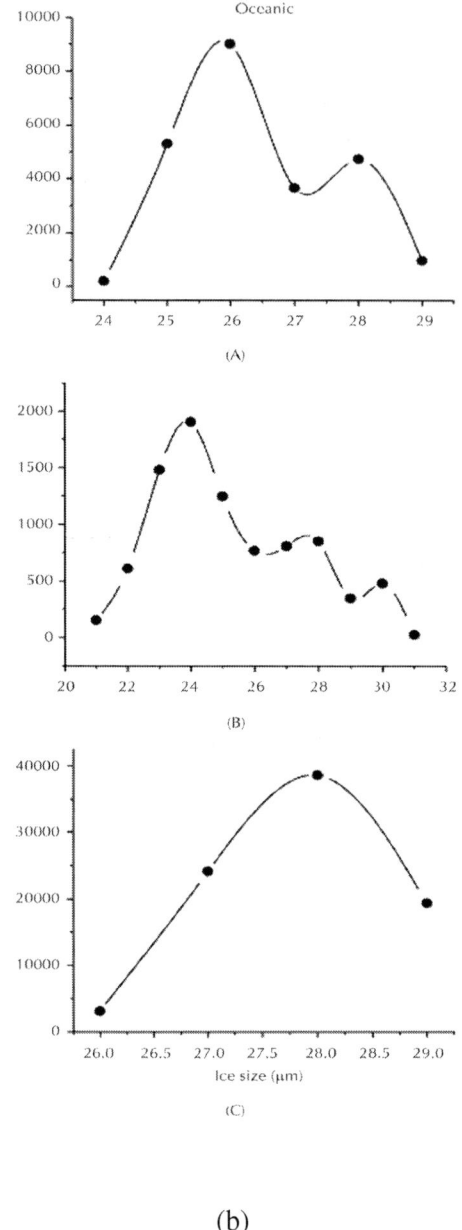

(b)

Figure 2: Relationship between mean cloud ice size and lightning over continental region (a) L1 (A), L2 (B), and L3 (C) and over oceanic region (b) O1 (A), O2 (B), and O3 (C).

In order to understand this relationship we have analyzed vertical distribution of cloud ice concentration and relationship between cloud size with ice concentration over the continental and oceanic region. Figures 3(a) and 3(b) show the cloud ice concentration at different altitude averaged during 2000–2011 period over the continental region L1 and oceanic region O2, respectively. It can be seen that ice concentration increases from altitude of 6 km, attends maximum concentration around 8–11 km over L1 and 10–14 km over O2 region, and decreases nearly to zero concentration at 18 km. Figure 4 shows the distribution of cloud ice concentration as a function of mean cloud size at an altitude of 12 km (near to same height) over L1 and O2 regions, respectively. It can be seen from Figure 4 that ice concentration over L1 and O2 regions increases with respect to ice size up to 24μm, attends maximum concentration at 24μ m, and ice concentration decreases with ice size above 24μ m. Similar relationship between ice concentration and ice size is also seen for the altitudes ranged between 8 and 14 km (not shown here). The charge is generated due to growth of ice size by condensation (deposition, combination of collection, etc.) and collision among them during upward motion [41, 43]. This process enhances the electric filed inside the cloud and generate lightning. During convection, cloud ice grows its size by combination of collection and condensation or deposition. The increasing ice concentration with respect to mean ice size from 19 to 24 μm in Figure 4 can be attributed to the growth of ice size. Therefore, generated charge (due to collision and condensation) increases with increasing the ice concentration and attends the maximum charge with the maximum ice concentration at 24μ m inside the cloud. Hence, increasing lightning frequency with increasing the ice concentration with respect to ice size from 19 to 24 μm with maximum lightning at 24μ m can be seen in Figure 2. Takahashi [27] has also found increase in lightning with increase in the ice concentration and ice size, compliments to our results. The latent heat is generated during condensation or deposition increases the updraft velocity of the hydrometers, which enhances the hydrometers concentration at high altitude (Figure 3) as well as electric filed inside the cloud. Ziegler and MacGorman [44] and Dey et al. [11] found that altitude range of about 7–10 km is favorable for electrification of clouds for generating lightning discharge. This is consistent with Figure 3 where we observed maximum ice concentration between 8 and 14 km altitude ranges. It can also be seen from Figure 3 that although maximum ice concentration over both land and oceanic regions are found approximately in same altitude range (8–14 km), yet less lightning occurs over oceanic region. It might be due to weak updraft velocity in mix-phase region over ocean as compared to continental region.

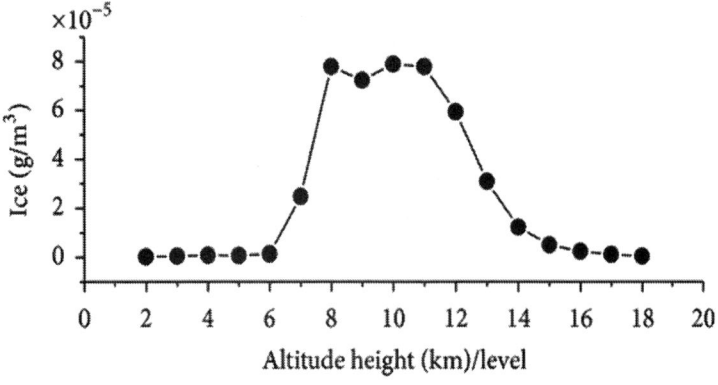

(a)

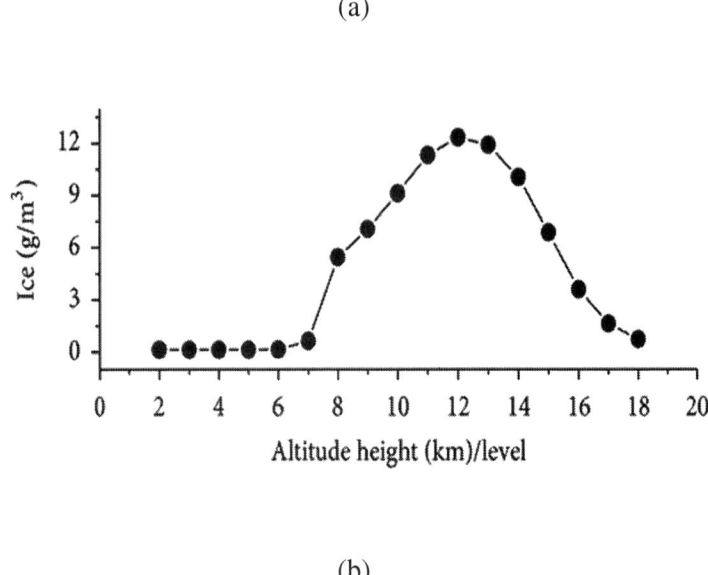

(b)

Figure 3: Distribution of cloud ice concentration as a function of altitude averaged during 2000–2011 period (a) over the continental region L1 and (b) oceanic region O2.

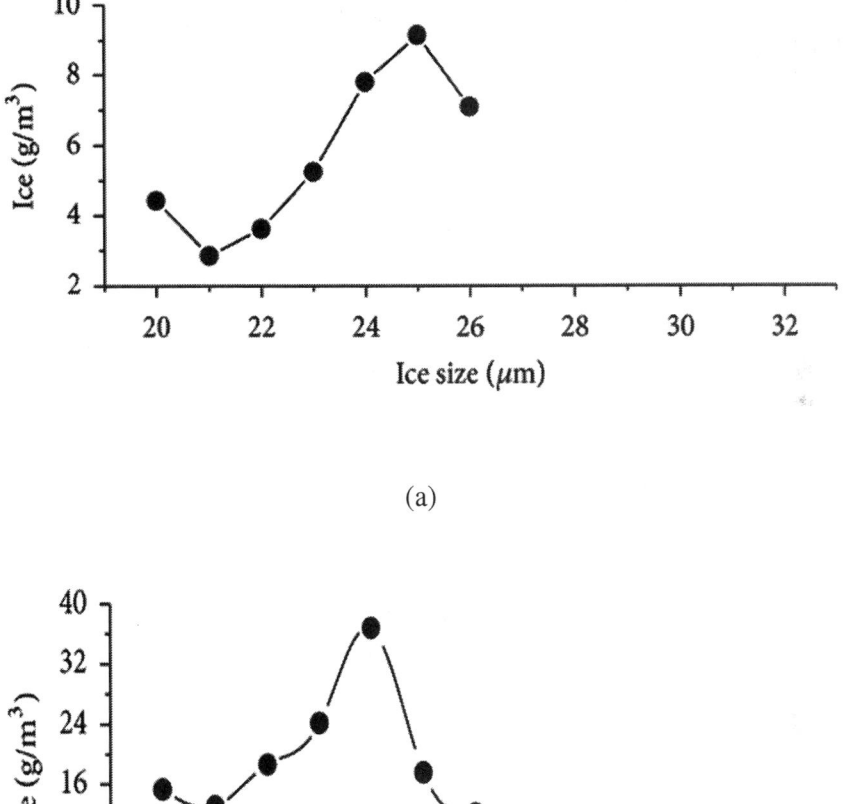

(a)

(b)

Figure 4: Relation between mean cloud ice size and cloud ice concentration averaged during 2000–2011 period at 12 km altitude (a) over the continental region L1 and (b) oceanic region O2.

On the other hand, increased size of cloud ice increases its terminal velocity and thereby reduces the uplift velocity. Therefore, larger particle descends isothermally towards the ground and begins to melt [45]. In this case charge with opposite polarity is also generated on hydrometer [43], however, unable to generate the charge to produce the lightning due to slow sublimation and low concentration of ice. Takahashi [4] also showed that the charge transfers per collision slow down with increase in ice diameter size. Therefore, the negative relationship between total lightning and mean cloud ice size which is greater than 26–28 µm instead of positive correlation for the size less than 19–25µ m can be seen in Figure 2. It is reasonable to conclude from Figures 2, 3, and 4 that the highest mean ice size of around 24µ m contributes to maximum ice concentration in the altitude range between 8 and 14 km and therefore results in the maximum lightning (over land and oceanic regions) for ice size of around 23–26 µm (seen in Figure 2). These imply that the relationship between mean ice size and lightning is curve linear.

CONCLUSIONS

In this work, we have analyzed 12 years (2000–2011) of monthly mean satellite observations of lightning from LIS, ice concentration from TRMM (3A12, V6), and effective diameter of cloud ice form MODIS over the Tropical Ocean and continental regions. We have examined the association of lightning flashes with mean ice size over these regions. A clear spatial change in lighting and cloud ice size from spring to winter season is seen. In general, total lightning is observed higher over the continental regions as compared to the lightning observed over oceanic region, whereas mean cloud ice size is observed higher over the oceanic region compared to the continental region during all the seasons. It is observed that the relationship between lightning and mean cloud ice size is same over both continental and oceanic regions. It is also observed that maximum lightning occurred for the mean cloud ice size of around 23–25 µm over the continental region and mean cloud ice size of around 24–28µ m over the oceanic region. However, for the first time, we found that relationship between lightning and mean cloud ice size follows the curve linear pattern and is not linear. We found that total lighting increases with increase in the cloud ice size and attends maximum at certain cloud ice size, then lightning decreases with increasing cloud ice size. The altitude profile show increase in ice concentration from 6 km, attends maximum concentration around 8–11 km over continent and 10–14 km over oceanic region, and decreases to zero concentration at around 18 km. Ice concentration within this region shows maximum around 24μm. This concludes that maximum lightning observed around 23–25 μm over the continental region and 24–28μ m over the oceanic region is associated with the large ice concentration at around 24μ m.

ACKNOWLEDGMENTS

The authors are gratefully thankful to Earth Observatory System and Earth System Science Program who provided the TRMM data. The authors also acknowledge the MODIS mission scientists and associated NASA personnel for the production of the data used in this research.

REFERENCES

1.   W. Deierling, W. A. Petersen, J. Latham, S. Ellis, and H. J. Christian, "The relationship between lightning activity and ice fluxes in thunderstorms," Journal of Geophysical Research: Atmospheres, vol. 113, no. D15, Article ID D15210, 2008.

2.   R. E. Newell, Y. Zhu, E. V. Browell, W. G. Read, and J. W. Waters, "Walker circulation and tropical upper tropospheric water vapor," Journal of Geophysical Research: Atmospheres, vol. 101, no. D1, pp. 1961–1974, 1996.   ·   ·

3.   S. E. Reynolds, M. Brook, and M. F. Gourley, "Thunderstorm charge separations," Journal of Meteorology, vol. 14, pp. 426–436, 1957.

4.   T. Takahashi, "Riming electrification as a charge generation mechanism in thunderstorms," Journal of the Atmospheric Sciences, vol. 35, pp. 1536–1548, 1978.

5.   D. G. Evans and W. C. A. Hutchinson, "The electrification of freezing water droplets and of colliding ice particles," Quarterly Journal of the Royal Meteorological Society, vol. 89, no. 381, pp. 370–375, 1963.   ·

6.   S. E. Reynolds, "Thunderstorm-precipitation growth and electrical-charge generation," Bulletin of the American Meteorological Society, vol. 34, pp. 117–123, 1953.

7.   P. R. Krehbiel, M. Brook, and R. A. McCrory, "An analysis of the charge structure of lightning discharges to ground," Journal of Geophysical Research: Oceans, vol. 84, no. 5, pp. 2432–2456, 1979.

8.   L. D. Carey and S. A. Rutledge, "A multiparameter radar case study of the microphysical and kinematic evolution of a lightning producing storm," Meteorology and Atmospheric Physics, vol. 59, no. 1-2, pp. 33–64, 1996.

9.   W. A. Petersen, S. A. Rutledge, and R. E. Orville, "Cloud-to-ground lightning observations from TOGA COARE: selected results and lightning location algorithms," Monthly Weather Review, vol. 124, no. 4, pp. 602–620, 1996.

10.   W. A. Petersen, S. A. Rutledge, R. C. Cifelli, B. S. Ferrier, and B. F. Smull, "Shipborne dual-Doppler operations during TOGA COARE: integrated observations of storm kinematics and electrification," Bulletin of the American Meteorological Society, vol. 80, no. 1, pp. 81–96, 1999.

11.   J. E. Dye, W. P. Winn, J. J. Jones, and D. W. Breed, "The electrification of New Mexico thunderstorms. 1. Relationship between precipitation development and the onset of electrification," Journal of Geophysical Research: Atmospheres, vol. 94, no. D6, pp. 8643–8656, 1989. · ·

12.   S. E. Reynolds and M. Brook, "Correlation of the initial electric field and the radar echo in thunderstorms," Journal of Meteorology, vol. 13, no. 4, pp. 376–380, 1956.

13.   E. R. Jayaratne, C. P. R. Saunders, and J. Hallett, "Laboratory studies of the charging of soft- hail during ice crystal interactions," Quarterly Journal of the Royal Meteorological Society, vol. 109, no. 461, pp. 609–630, 1983.

14.   E. R. Williams, "The tripole structure of thunderstorms," Journal of Geophysical Research: Atmospheres, vol. 94, no. D11, pp. 13151–13167, 1989.

15.   C. P. R. Saunders and S. L. Peck, "Laboratory studies of the influence of the rime accretion rate on charge transfer during

crystal/graupel collisions," Journal of Geophysical Research: Atmospheres, vol. 103, no. D12, pp. 13949–13956, 1998.

16. E. R. Mansell, D. R. MacGorman, C. L. Ziegler, and J. M. Straka, "Charge structure and lightning sensitivity in a simulated multicell thunderstorm," Journal of Geophysical Research: Atmospheres, vol. 110, no. D12, pp. 1–24, 2005. · ·

17. J. Latham, W. A. Petersen, W. Deierling, and H. J. Christian, "Field identification of a unique globally dominant mechanism of thunderstorm electrification," Quarterly Journal of the Royal Meteorological Society, vol. 133, no. 627, pp. 1453–1457, 2007. · ·

18. E. J. Zipser, "Deep cumulonimbus cloud systems in the tropics with and without lightning," Monthly Weather Review, vol. 122, no. 8, pp. 1837–1851, 1994.

19. E. R. Williams, R. Zhang, and J. Rydock, "Mixed-phase microphysics and cloud electrification," Journal of the Atmospheric Sciences, vol. 48, no. 19, pp. 2195–2203, 1991.

20. R. G. Pereyra, E. E. Avila, N. E. Castellano, and C. P. R. Saunders, "A laboratory study of graupel charging," Journal of Geophysical Research: Atmospheres, vol. 105, no. D16, pp. 20803–20812, 2000.

21. C. P. R. Saunders, H. Bax-Norman, C. Emersic, E. E. Avila, and N. E. Castellano, "Laboratory studies of the effect of cloud conditions on graupel/crystal charge transfer in thunderstorm electrification," Quarterly Journal of the Royal Meteorological Society, vol. 132, no. 621, pp. 2653–2673, 2006. · ·

22. V. N. Bringi, I. J. Caylor, J. Turk, and L. Liu, "Microphysical and electrical evolution of a convective storm using multiparameter radar and aircraft data during CaPE," in Proceedings of the 26th International Conference on Radar Meteorology, pp. 312–314, American Meteorological Society, May 1993.

23. J. E. Dye, "Early electrification and precipitation development in a small, isolated Montana cumulonimbus," Journal of Geophysical Research: Atmospheres, vol. 91, no. D1, pp. 1231–1247, 1986.

24. J. E. Dinger and R. Gunn, "Electrical effects associated with a change of state of water," Terrestrial Magnetism and Atmospheric Electricity, vol. 51, no. 4, pp. 477–494, 1946. ·

25. M. Stolzenburg, T. C. Marshall, W. D. Rust, and B. F. Smull, "Horizontal distribution of electrical and meteorological conditions across the stratiform region of a mesoscale convective system," Monthly Weather Review, vol. 122, no. 8, pp. 1777–1797, 1994.

26. D. Rosenfeld, U. Lohmann, G. B. Raga et al., "Flood or drought: how do aerosols affect precipitation?"Science, vol. 321, no. 5894, pp. 1309–1313, 2008. · ·

27. T. Takahashi, "Thunderstorm electrification—a numerical study," Journal of the Atmospheric Sciences, vol. 41, no. 17, pp. 2541–2558, 1984.

28. S. C. Sherwood, V. T. J. Phillips, and J. S. Wettlaufer, "Small ice crystals and the climatology of lightning," Geophysical Research Letters, vol. 33, no. 5, Article ID L05804, 2006. ·

29. S. C. Sherwood, "Aerosols and ice particle size in tropical cumulonimbus," Journal of Climate, vol. 15, no. 9, pp. 1051–1063, 2002.

30. D. M. Lal and S. D. Pawar, "Relationship between rainfall and lightning over central Indian region in monsoon and premonsoon seasons," Atmospheric Research, vol. 92, no. 4, pp. 402–410, 2009. · ·

31. S. W. Nesbitt, E. J. Zipser, and D. J. Cecil, "A census of precipitation features in the tropics using TRMM: radar, ice scattering, and lightning observations," Journal of Climate, vol. 13, no. 23, pp. 4087–4106, 2000.

32. W. Deierling, J. Latham, W. A. Petersen, S. M. Ellis, and H. J. Christian Jr., "On the relationship of thunderstorm ice hydrometeor characteristics and total lightning measurements," Atmospheric Research, vol. 76, no. 1–4, pp. 114–126, 2005. · ·

33. W. A. Petersen, H. J. Christian, and S. A. Rutledge, "TRMM observations of the global relationship between ice water content and lightning," Geophysical Research Letters, vol. 32, no. 14, Article ID L14819, 2005. ·

34. K. C. Wiens, S. A. Rutledge, and S. A. Tessendorf, "The 29 June 2000 supercell observed during STEPS—part II: lightning and charge structure," Journal of the Atmospheric Sciences, vol. 62, no. 12, pp. 4151–4177, 2005. · ·

35. D. M. Lal and S. D. Pawar, "Effect of urbanization on lightning over four metropolitan cities of India,"Atmospheric Environment, vol. 45, no. 1, pp. 191–196, 2011. · ·

36. S. M. Steiger, R. E. Orville, and G. Huffines, "Cloud-to-ground lightning characteristics over Houston, Texas: 1989–2000," Journal of Geophysical Research: Atmospheres, vol. 107, no. D11, pp. 2–13, 2002. · ·

37. E. J. Workman and S. E. Reynolds, "Electrical activity as related to thunderstorm cell growth," Bulletin of the American Meteorological Society, vol. 30, pp. 142–149, 1949.

38. E. R. Williams and R. M. Lhermitte, "Radar tests of the precipitation hypothesis for thunderstorm electrification," Journal of Geophysical Research: Oceans, vol. 88, no. C15, pp. 10984–10992, 1983.

39. S. A. Rutledge, E. R. Williams, and T. D. Keenan, "The Down Under Doppler and Electricity Experiment (DUNDEE): overview and preliminary results," Bulletin of the American Meteorological Society, vol. 73, no. 1, pp. 3–16, 1992.

40. R. A. Black, J. Hellett, and C. R. P. Saunders, "Air craft study of precipitation and electrification," inProceeding of the 17th Conference on Severe Local Storms and Conference on Atmospheric Electricity, pp. J20–J25, St. Louis, Mo, USA, October 1993.

41. H. J. Christian, R. J. Blakeslee, S. J. Goodman, et al., "The lightning imaging sensor," in Proceedings of the 11th International Conference on Atmospheric Electricity, pp. 746–749, Guntersville, Ala, USA, June 1999.

42. Annual Climate Summary, IMD Government of India, 2005.

43. J. P. Rydock and E. R. Williams, "Charge separation associated with frost growth," Quarterly Journal of the Royal Meteorological Society, vol. 117, no. 498, pp. 409–420, 1991.

44. C. L. Ziegler and D. R. Macgorman, "Observed lightning morphology relative to modeled space charge and electric field distributions in a tornadic storm," Journal of the Atmospheric Sciences, vol. 51, no. 6, pp. 833–851, 1994.

45. T. R. Shepherd, W. D. Rust, and T. C. Marshall, "Electric fields and charges near 0°C in stratiform clouds," Monthly Weather Review, vol. 124, no. 5, pp. 919–938, 1996.

Chapter 3

The West African Sahel: A Review of Recent Studies on the Rainfall Regime and Its Interannual Variability

Sharon E. Nicholson

Earth, Ocean, and Atmospheric Sciences Department, Florida State University, Tallahassee, FL 32306, USA

ABSTRACT

The West African Sahel is well known for the severe droughts that ravaged the region in the 1970s and 1980s. Meteorological research on the region has flourished during the last decade as a result of several major field experiments. This paper provides an overview of the results that have ensued. A major focus has been on the West African monsoon, a phenomenon that links all of West Africa. The characteristics and revised picture of the West African monsoon are emphasized. Other .

topics include the interannual variability of rainfall, the atmospheric circulation systems that govern interannual variability, characteristics of precipitation and convection, wave activity, large-scale factors in variability (including sea-surface temperatures), and land-atmosphere relationships. New paradigms for the monsoon and associated ITCZ and for interannual variability have emerged. These emphasize features in the upper atmosphere, as well as the Saharan Heat Low. Feedback mechanisms have also been emphasized, especially the coupling of convection with atmospheric dynamics and with land surface characteristics. New results also include the contrast between the premonsoon and peak monsoon seasons, two preferred modes of interannual variability (a latitudinal displacement of the tropical rainbelt versus changes in its intensity), and the critical importance of the Tropical Easterly Jet.

INTRODUCTION

The Sahel region of West African is a semi-arid expanse of grassland, shrubs, and small, thorny trees lying just to the south of the Sahara desert (Figure 1). The term is often applied to the general region extending some 5000 km across the east-west extent of Africa and from the Sahara to the humid savanna at roughly 10° North. "Sahel" more properly applies to a smaller region (Figure 2) between the latitudes of roughly 14° N and 18° N. It includes much of the countries of Mauritania, Senegal, Mali, Niger, Chad, the Sudan, and the northern fringes of Burkina Faso and Nigeria. The Sahel's highly diverse inhabitants have a long history, its fabled cities such as Timbuktoo and Djenne having prospered as centers of trade, education, and political empires many centuries ago. Today the region is home to major cities such as Dakar (Senegal), Niamey (Niger), Bamako (Mali), and Khartoum (Sudan), but most of the inhabitants live in rural areas and practice agriculture. Hence, the vagaries of climate are of great importance to the region.

Figure 1: Typical low tree and shrub savanna landscape of the Sahel (from [57]).

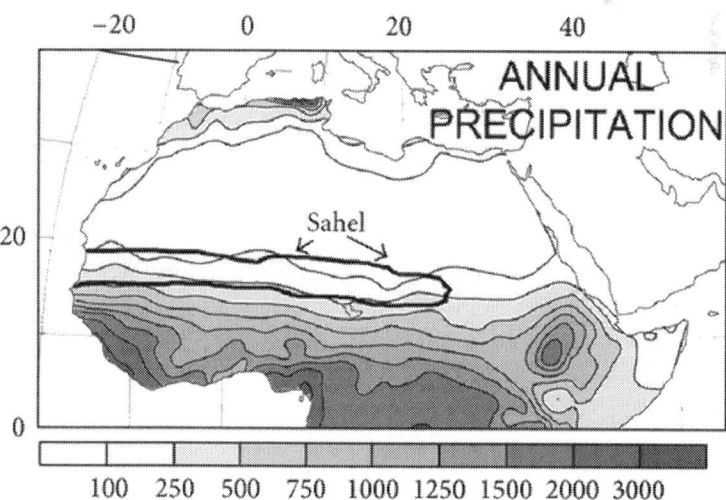

Figure 2: Mean annual precipitation over West Africa (in mm). Location of the Sahel is indicated.

Mean annual rainfall in the Sahel is on the order of 100 to 200 mm in the north, where the Sahel gives way to desert, and 500 to 600 mm at its southern limit (Figure 2). Throughout the region rainfall is generally limited to the boreal summer months, with maximum rainfall occurring in August. In the Sahel proper, the season length ranges from 1 to 2 months in the north to 4 to 5 months in the south. Occasional winter rains of extratropical origin can occur, but these generally bring less than 25 mm of rainfall.

Within the region there is a strong east-west uniformity of climate and vegetation conditions. This is well illustrated by the isohyets of rainfall (Figure 2), which also show a very strong north-south gradient. Because of the east-west uniformity the Sahel region is often considered as one entity in the meteorological context. However, notable contrasts appear across the region, particularly with respect to factors in year-to-year variability and the prevailing circulation systems [1, 2]. This is particularly true in the eastern extreme of the Sudan and Ethiopia, where complex topography overrides the large-scale patterns. For this reason, most analyses of Sahel climate do not extend across the region's east-west extent. Those of Nicholson (e.g., [3]) generally cease at 30° to 35° E. Those of Lamb (e.g., [4, 5]) are confined to the region lying between the Atlantic coast and 10° E. Several studies have shown significant contrasts in various sectors of the Sahel (e.g., [6–8]).

Scientific attention was focused on the Sahel in the 1970s and 1980s because of the long period of drought that had ravaged the region and the controversial issue of desertification. During the past few years the Sahel once again received much meteorological attention, at least in part because of major field experiments carried out to better understand variability in the region. These include the AMMA (African monsoon multidisciplinary analysis) experiment, that took place in 2006 [9, 10], the associated model intercomparison project (ALMIP, [11]), the AMMA Catch Experiment [12], which extended AMMA southward into Benin, and the JET2000 Experiment, that focused on the African Easterly Jet [13].

This paper presents a review of studies that have been published in roughly the last decade, a period during which our meteorological understanding of the region increased tremendously. The conclusions within this paper are generally valid for the broader region termed "Sahel", from the Sahara to the more humid savanna, and from the far west to roughly 30° E.

The Sahel is inextricably linked to the West Africa monsoon, the circulation regime that brings most of its rainfall. This paper therefore begins with an overview of the concept of the West African monsoon and the new picture that has emerged in recent years. The paper focuses on interannual and intraseasonal variability in the Sahel and factors governing this variability, including links to large-scale phenomena such as ENSO, SSTs, and atmospheric circulation. The impact of aerosols and feedback from the Sahelian land surface are also considered. The paper further describes new characteristics that have been demonstrated about the region's storm systems and new circulation features that have been documented. The paper is limited to precipitation and related phenomena.

THE WEST AFRICAN MONSOON

Classic Picture of the West African Monsoon

The classical picture of the West African monsoon is illustrated in Figure 3. In this scenario, rainfall is associated with a surface feature termed the Intertropical Convergence Zone or ITCZ. This zone "follows the sun", moving northward into West Africa in the boreal summer and southward into southern Africa in the austral summer, twice traversing the equatorial regions. In this classic picture, the ITCZ over West Africa is marked by the convergence of the northeasterly Harmattan winds that originate in the Sahara and the southwest monsoon flow that emanates from the Atlantic. Rain production was assumed to result from local thermal instability, facilitated by the low-level wind convergence within this zone. The rapid increase in rainfall from the Sahara to the humid equatorial zone was assumed to relate to a rapidly increasing depth of the moist layer equator-ward from the ITCZ (e.g, [14]).

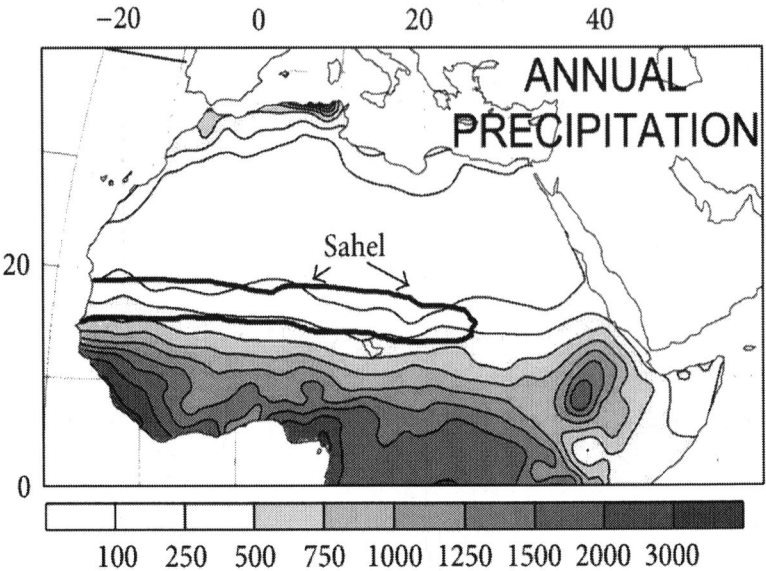

Figure 3: Classical picture of the ITCZ over Africa during the boreal summer (from [14]).

This picture of the West African monsoon has several shortcomings. For one, its origin lies in the concept of the global Hadley circulation, with rising motion in the equatorial latitudes where the trade winds converge and subsidence (sinking motion) in the subtropical latitudes where the subtropical highs prevail. That concept was developed to describe the global mean state and not individual regions. Moreover, it is primarily valid over the oceans [15], where the trade winds are well developed. The scheme works poorly over the continents, where the trade winds are generally absent. This picture of the ITCZ was also developed at a time when tropical rainfall was assumed to be local in origin, a result of thermal instability in warm and humid air, with ascent facilitated by surface wind convergence. As of the 1970s, following tropical meteorological experiments such as GATE (the GARP Atlantic Tropical Experiment, [16]), it became well established that tropical rainfall is instead linked to large-scale disturbances often associated with traveling waves.

A second shortcoming is the ambiguity of the term "ITCZ". In the literature three very different definitions are given, based respectively on wind convergence, surface air pressure and rainfall or outgoing longwave radiation. In the Encyclopedia of World Climatology, Yan and Oliver [17] states that the ITCZ is "an east-west oriented low-pressure region near the equator where surface northeasterly and southeasterly trade winds meets. When they converge, moist air is forced upward, producing cumulus clouds and heavy precipitation." According to Miller and Schneider [18] in the Encyclopedia of Weather and Climate, it is "a region near the equator where the trade winds converge." Holton et al. [19], in their widely used textbook, define the ITCZ as the "loci of cloud clusters associated with westward-propagating tropical wave disturbances". Consequently, some authors now avoid the use of the term. For example, Zhang et al. [15] instead use the term "rain band" and Nicholson [20] substitutes the term "tropical rainbelt". The latter term will be used in this paper.

As a result of the ambiguity in the definition of the ITCZ, its tracking may be based on a pressure minimum, a surface wind convergence, a maximum in rainfall, a minimum in outgoing longwave radiation, or a maximum in cloudiness. The use of so many different parameters has been justified by the assumptions that (1) the pressure minimum and rainfall maximum are colocated with each other and with the wind convergence, (2) maximum cloudiness is roughly colocated with maximum rainfall, and (3) longwave radiation is at a minimum at that location. Unfortunately, these assumptions, especially the first one, do not stand up to close scrutiny. Even over the ocean regions the zone of minimum pressure does not generally coincide with that of the wind convergence or the rainfall maximum [21].

This picture is particularly problematic over West Africa, as new research on the West African monsoon has dramatically shown. For one, the ITCZ (as defined by surface wind convergence) lies some 1000 kilometers to the north of the zone of maximum rainfall, as does the zone of maximum ascent [20]. For another, rainfall over the Sahel is associated with African easterly waves and with large Mesoscale Convective Systems (MCSs), rather than local thunderstorms. Lebel et al. [22] estimate that some 12% of the total number of MCSs produce 90% of the rainfall during the peak rainy season. Very intense MCSs, which comprise some 3 to 4% of the all rain events, produce up to 80% of the rainfall that occurs in the Sahel [23].

Revised View of the West African Monsoon

The circulation over West Africa exhibits the most basic characteristics of a monsoon: a pronounced seasonal wind shift that is produced by thermodynamic contrasts between the land (i.e., the Sahara) and ocean (i.e., the equatorial Atlantic). Southwesterly flow is established between the Atlantic cold tongue (cool water close to the equator between the boreal spring and summer) and the Saharan heat low, bringing moisture into the continent [24, 25]. The annual evolution of moisture fluxes, convergence and rainfall is closely tied to these two systems. The basic surface circulation is illustrated in Figure 4. During the boreal summer, an intense heat low develops over the Western Sahara. The cyclonic flow around this low includes the southwesterly "monsoon" flow to the south and the northeasterly Harmattan to the west of its core [26]. Termed the Saharan Heat Low or West African Heat Low, this system plays an important role in controlling the northward penetration of the monsoon [27].

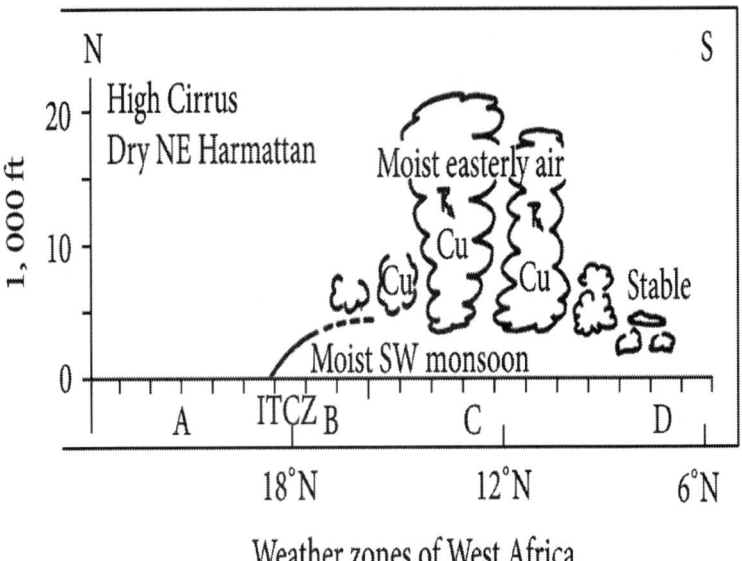

Weather zones of West Africa.

(a)

January circulation

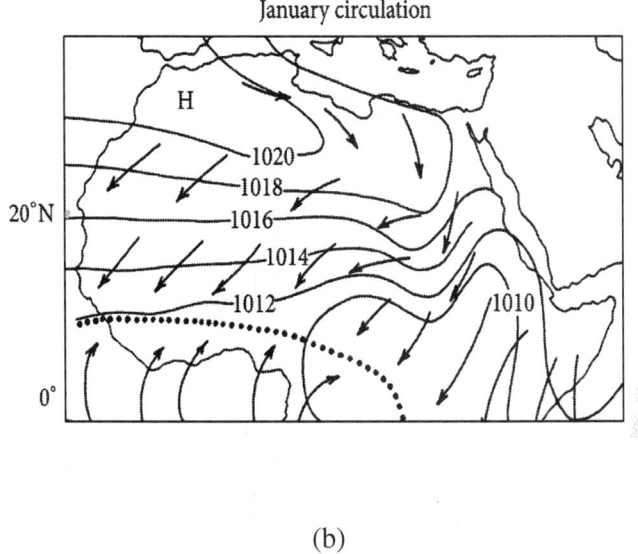

(b)

July/august circulation

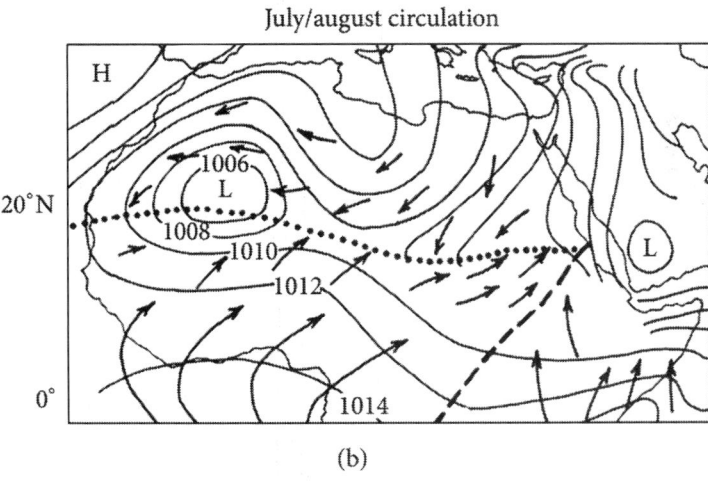

(b)

(c)

Figure 4: Schematic of surface wind (arrows) and pressure (mb) over West Africa during winter and at the peak of the summer monsoon.

The main contrasts with the classic picture of the monsoon are the diminished importance of the ITCZ and the inclusion of several jetstreams and shear zones, the African Easterly Waves (AEWs), the Saharan Heat Low, and the aforementioned Mesoscale Convective Systems, as opposed to local rainfall induced by thermal instability. Overviews are presented by Nicholson and Grist [1], Gu and Adler [28], Parker et al. [29], Zhang et al. [15], Nicholson [20], and Thorncroft et al. [25]. The circulation and convective features associated with the West African monsoon and influencing the Sahel are discussed in detail in Sections 4and 5.

The main tropical circulation features associated with the West African monsoon (Figures 5(a) and 5(b)) are the upper-level Tropical Easterly Jet (TEJ), the mid-level African Easterly Jet (AEJ), and low-level equatorial westerlies associated with the southwest monsoon flow [1]. In wet years these westerlies become a bona fide jet stream that has a core near 850 mb and is independent of the low-level monsoon flow [30]. The AEJ and monsoon westerlies are stronger over the western portion of the region, while the TEJ is stronger over the eastern portion of the region (Figure 5(b)). Superimposed upon these zonal flows are two meridional overturning circulations: a deep circulation associated with low-level contrasts in deep moist convection and a shallow circulation associated with contrasts in dry convection [25] (Figure 6). Both consist of southerly flow at low levels and northerly flow at higher levels. The shallow cell is associated with the Saharan heat low and transports dry air towards the rainbelt at mid-levels of the atmosphere [20].

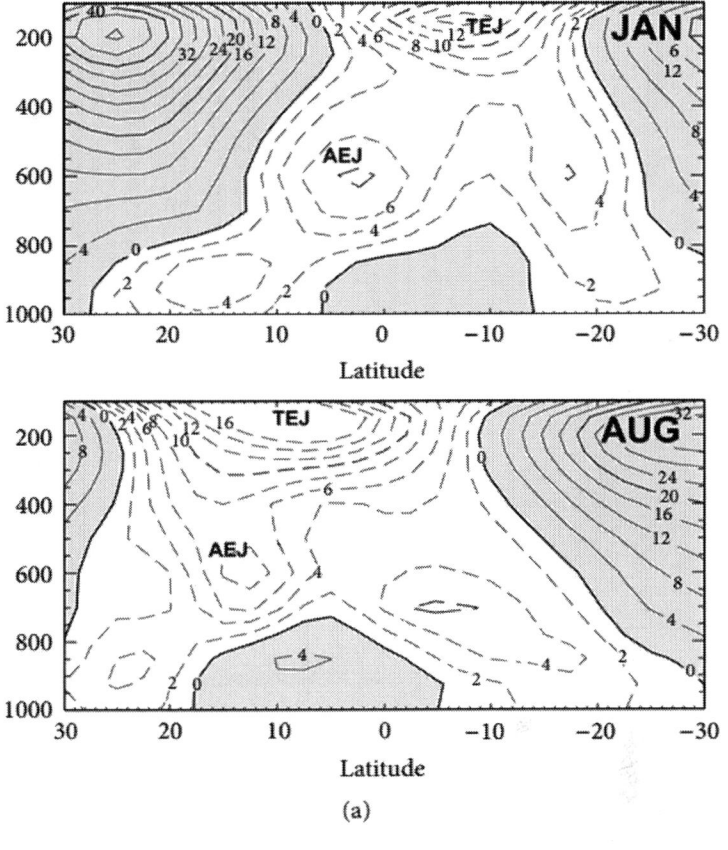

(a)

(a)

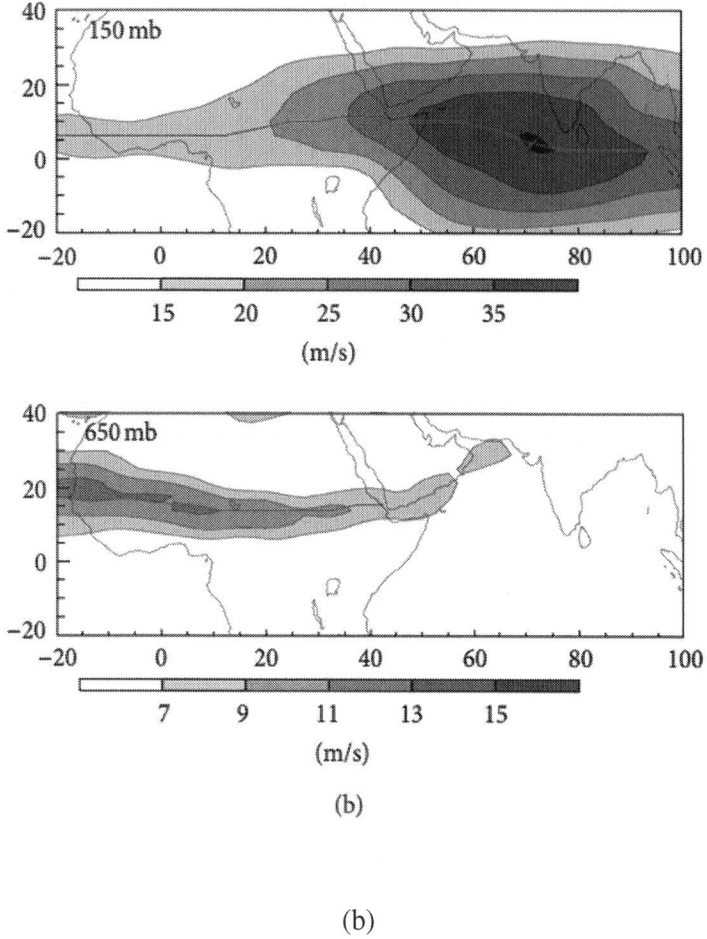

(b)

Figure 5: (a) Latitudinal transect of mean zonal wind (m s^{-1}) over Africa during January and August (from [57]). The Tropical Easterly Jet and African Easterly Jet are indicated. Westerly winds are shaded. (b) Mean easterly wind speed (m s^{-1}) in August at 600 and 150 hPa, showing the African Easterly Jet and Tropical Easterly Jet, respectively (from [20]).

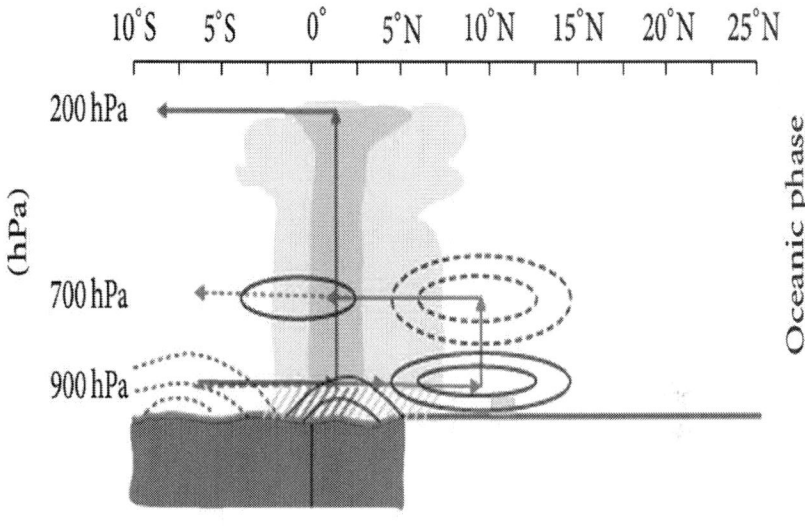

(a)

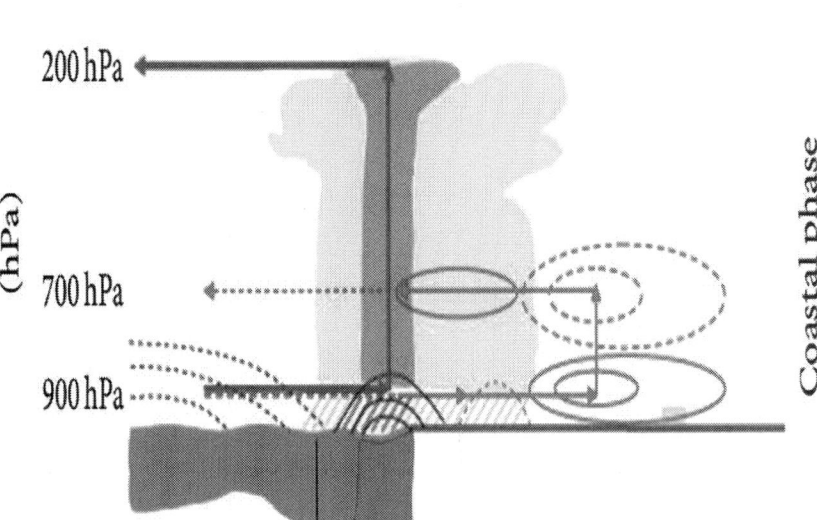

(b)

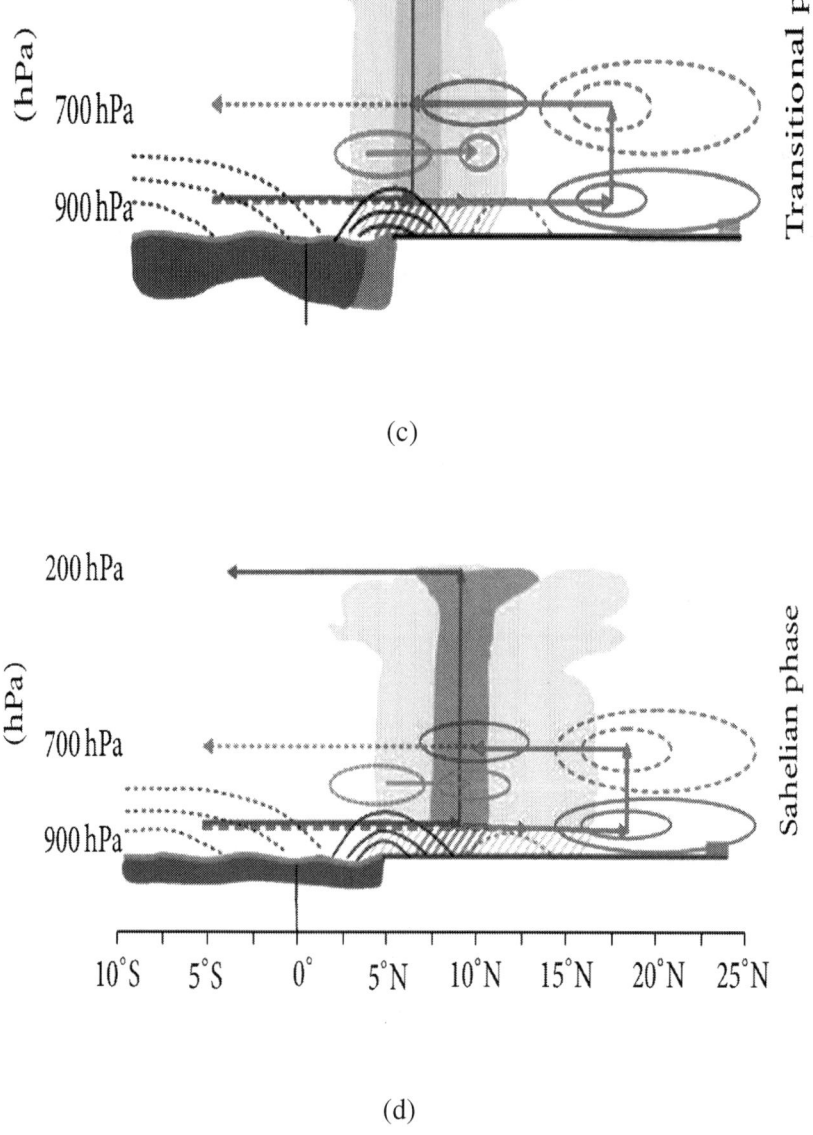

Figure 6: Schematic showing the four key phases of the annual cycle of the west African monsoon. Included for each phase are the following: location

of the main rain band (indicated by clouds and rainfall with peak values highlighted by darker shaded clouds and rainfall), the location of the Saharan heat-low (indicated by yellow, orange, and red shading at the surface pole-ward of the rain band, with increased redness indicating increased intensity). Atlantic ocean temperature and associated mix-layer depth (with decreased temperatures indicated by the red-to-green-to-blue transition). Moisture flux convergence maxima and minima (solid contours indicate moisture flux convergence and dashed contours indicate moisture flux divergence), and the deep and shallow meridional circulations (blue and red lines with arrows); dashed lines suggest some uncertainty about the extent to which Shallow Meridional Circulation return flow penetrates the latitude of the main rain band or not. The moisture flux convergence quadrupole structure is high-lighted by red contours and the dipole at 850 hPa structure is highlighted by green contours (from [25]).

The core of the AEJ is between 650 and 700 mb. It lies in a region of strong latitudinal temperature gradient in the lower troposphere. The jet is thermally induced by the contrast between the hot Sahara and the Atlantic Ocean, but maintained by the juxtaposition of moist convection to the south and dry convection to the north [31–33]. It is also associated with zones of intense horizontal and vertical shear [1, 28]. The horizontal shear is particularly strong in August and September, while the vertical shear is most intense from May through July, when it is of roughly the same magnitude as the horizontal shear.

The TEJ is best developed at about the 200 mb level. It has a sizable meridional component [30] that is part of the deep meridional overturning described by Thorncroft et al. [25]. Over West Africa during the monsoon season, the meridional component is northerly south of its core (maximum is near 10 degrees South), but southerly north of the core. This creates a region of strong divergence around 200 mb, which appears to play a role in the development of the rainbelt over West Africa [1, 20]. It intensifies during the monsoon season.

A major role of the West African monsoon system is transporting moisture into West Africa from the Atlantic. This transport takes place in periodic northward excursions of moisture flux that have a 3 to 5 day time scale [34]. These "bursts" of the monsoon can bring copious rainfall to the northern fringes of the Sahel [35]. The intraseasonal excursions in moisture flux can be stationary or exhibit a westward propagation [34]. The dynamics of the Saharan Heat Low are a major driver in the stationary perturbations. They follow maxima in the intensity of the heat low and the associated acceleration of the low-level meridional wind. The propagating excursions of moisture flux are related to easterly waves. Horizontal advection is the main process in the moisture flux, but vertical turbulent exchange plays some role.

Within the monsoon are two areas of peak moisture flux convergence and vertical motion [25, 36]. The dominant one lies between the axes of the AEJ and TEJ (Figure 7). It is associated with the deep meridional overturning described by Thorncroft et al. [25] and with the core of the rainbelt over West Africa [1]. The mean August location of this deep column of ascent is roughly 10° North. This region of moisture flux convergence and ascent is associated with a deep column of moist air in which relative humidity is 60% to 80% throughout the troposphere [20, 37]. It also corresponds to the southerly track of African Easterly Waves (Figure 8). Notably, at low levels both the moisture convergence and the vertical motion peak at the West African coast. Frictional uplift when the low-level southwesterlies meet the land surface probably plays some role.

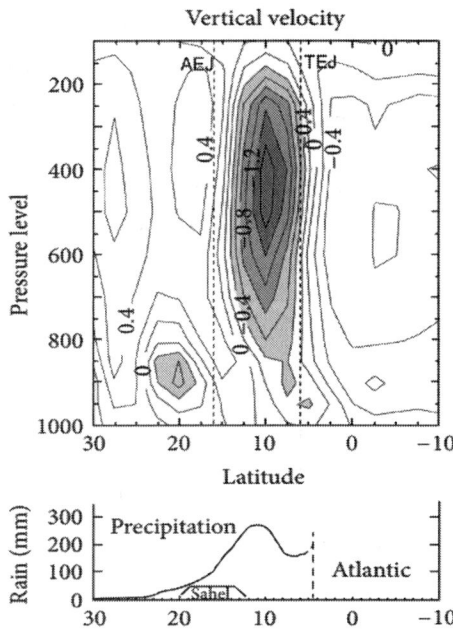

Figure 7: Schematic of the rainbelt over West Africa (from [20]). Top diagram is a vertical cross-section of mean vertical motion (10^{-2} hPas^{-1}) in August. The

main region of ascent lies between the axes of the African easterly jet (AEJ) and the tropical easterly jet (TEJ). A shallow region of ascent corresponds to the surface position of the Intertropical Convergence Zone (ITCZ) and the center of the Saharan heat low. The bottom diagram gives mean August rainfall (mm mo^{-1}, averaged for 10° W to 10° E) as a function of latitude, with the location of the Sahel indicated on the latitudinal axis.

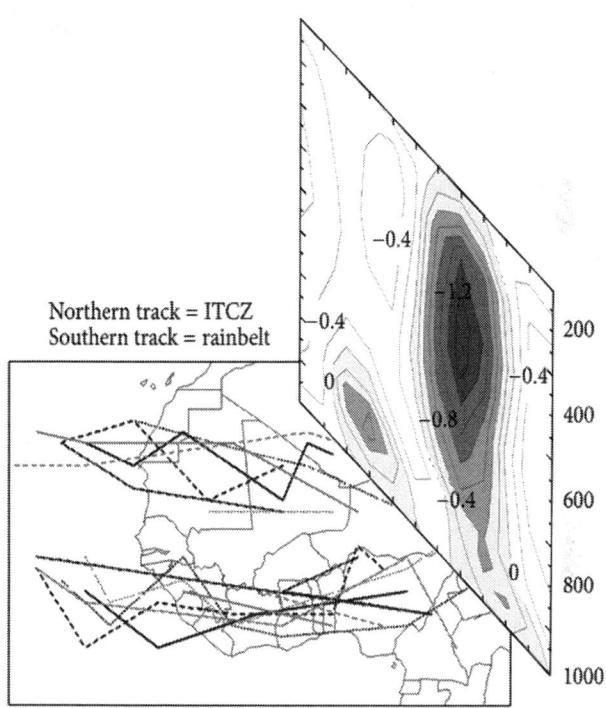

Figure 8: The relationship between tracks of African Easterly Waves and the vertical motion fields (omega) over West Africa (based on [36]). The tracks are illustrated using waves of August 1982.

A second area of ascent and moisture flux convergence is associated with the heat low over the Sahara [20,25]. It is confined to low levels (Figures 6 and 9) and is part of the shallow meridional overturning described by Nolan et al. [32] and Zhang et al. [33]. This second region of ascent corresponds to the northerly track of African Easterly Waves, but waves in this track seldom produce rainfall. This region lies some 8 degrees poleward of the rainfall maximum. It also coincides with the surface manifestation of the ITCZ, that is, the convergence of the southwesterly monsoon flow and the northeasterly Harmattan. The ITCZ and the rainfall maximum are separated by a region of subsidence (Figures 7 and 9), indicating that the ITCZ is effectively "decoupled" from the rainbelt. The Sahel lies, on average, in this region of subsidence. The two regions of ascent merge in the wettest years in the Sahel [36].

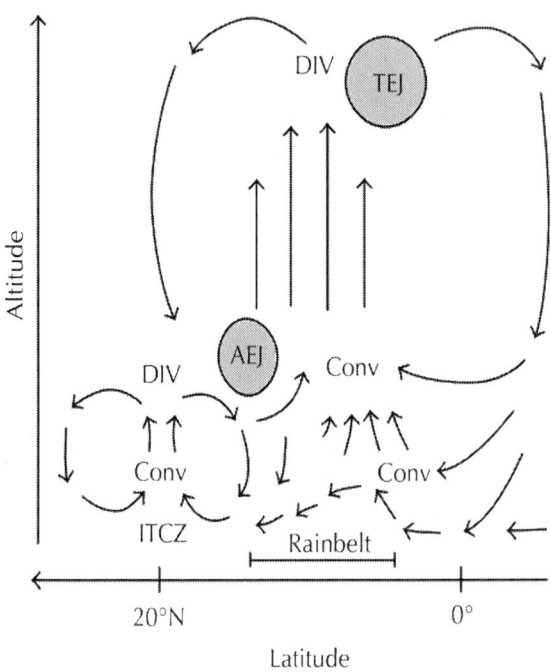

Figure 9: Schematic illustration of the revised picture of the West African monsoon (from [20]).

The Monsoon Onset

Another topic that has received much attention in the last decade is the onset of the West African monsoon, including its predictability (e.g., [38]) and association with atmospheric dynamics [39]. A difficulty has been in the approach to defining this onset and several varied parameters have been utilized to do so. Some have defined it in terms of simple rainfall characteristics. Marteau et al. [40], for example, use threshold values of daily rainfall at individual stations to quantify the magnitude and length of the initial wet spell and any subsequent breaks.

The mean onset date over the Sahel is June 24 [41, 42]. The onset proceeds slowly and is characterized by a succession of active and inactive phases (breaks in the monsoon) [43]. There is little spatial coherence in the onset dates and, consequently, little predictability [40]. The lack of coherence is a result of the fragmented nature of the active and break phases [44]. The withdrawal phase is relatively abrupt and rather uniformly distributed throughout the entire monsoon region.

Model simulations of Flaounas et al. [45] suggest that large-scale dynamics, extending beyond West Africa, control the monsoon onset and its timing. The onset may be related to the Indian monsoon [46, 47]. The associated convection north of 15° N forces a westward propagating Rossby wave that reaches West Africa some 7 to 15 days later, with the wave passage triggering convection in the Sahel.

Drobinski et al. [42] examine a case of late monsoon onset in Niger and Mali and demonstrated additional factors. During 2006 the onset there was delayed until July 3. The delay was associated with an anomalous direction of the AEJ, which flowed northeasterly in a band over the Hoggar and Aír Mountains. There was also an unusually strong northeast Harmattan in the lee of the mountains. The delay appeared to be related to the interaction of the AEJ and orography.

Rainfall Distribution in the West African Monsoon

Thorncroft et al. [25] define four phases of the West African monsoon, depending on the location of the rainfall peak: oceanic, coastal, transitional, and Sahelian. During the oceanic phase, between November and mid-April, a broad rainbelt lies just north of the equator. During the subsequent coastal phase, which generally prevails to mid-June, peak rainfall lies over the ocean but in the near-coastal region around 4 to 5° North [25, 28, 48–50]. The transition phase, when a decrease in rainfall is observed, occurs in early July. Lebel et al. [22] refer to these first three phases collectively as the oceanic regime. The

Sahelian phase lasts from mid-July to September. Throughout this phase the rainfall peak is more intense and remains just to the south of the Sahel, around 10° North. Rainfall in the Sahel is associated with this maximum, which Lebel et al. [22] term the continental regime.

The shift between the maximum at 5° North in the coastal phase and 10° North in the continental phase is very abrupt [41, 49], evoking the term "monsoon jump". Several authors have proposed explanations for the abrupt shift. Sultan and Janicot [41] suggest it is triggered by westward-propagating disturbances. Sijikumar et al. [51] and Ramel et al. [52] implicate the Saharan heat low, which intensifies and shifts northward at the time of the "jump". Gu and Adler [28] point out that the shift is associated with a northward shift of the African Easterly Jet and associated horizontal and vertical shear zones, as well as the development of westward-propagating waves. Okumura and Xie [53] suggest that the shift is related to the interaction of the Atlantic equatorial cold tongue and the African monsoon. Sultan et al. [50] suggest that complex interactions among convection, AEJ dynamics and local topography, especially the Ahaggar Plateau and Tibesti highlands, play a role. They also point out that when the heat low is sufficiently intense, the result is a reversal in the potential vorticity gradient and, consequently, the generation of AEWs and convection. Accordingly, the shift commences with a release of potential instability. Then inertial instability shifts the rain band to 10° N [54].

What underlies all of these mechanisms is a northward shift and intensification of the latitudinal temperature and pressure gradients over West Africa. The changes in the heat low and the instability mechanisms are probably direct consequences of the increased gradients. The northward shift of the AEJ, shear zones, waves disturbances, and convection can be viewed as results of the aforementioned factors.

There are notable contrasts between the characteristics associated with the two spatial rainfall maxima. One is in the origin of precipitation [28]. Early in the rainy season (May-June), when the maximum is near the Gulf of Guinea, rain-bearing synoptic systems tend to be eastward-propagating wave signals. In the late rainy season, when the maximum lies well into the continental interior, westward-propagating wave signals (i.e., the AEWs) are the dominant rain-bearing synoptic systems. The convective rain rate and the percent of area covered by cumulonimbus anvils are lower and the stratiform fraction of rainfall is higher in the second spatial maximum (which receives the bulk of its rain during the peak monsoon season) [48, 55]. Consistent with this, the size and organization of convective systems is greater during the peak monsoon season, but rainfall intensity is lower than earlier in the season [56].

The two spatial maxima also differ in the patterns of variability in recent years, with the second maximum (associated with the continental regime) exhibiting much more change. The rainfall peak associated with this regime has also appeared increasingly early in the season [58], with the August peak disappearing in recent years [8].

Intraseasonal Variability of the Monsoon

Most recent research on the West African monsoon's intraseasonal variability (time scale of 10 to 90 days) has centered on the Madden-Julian Oscillation or MJO. Its origin is generally over the Pacific warm pool, from which an eastward-propagating Kelvin wave and a westward-propagating Rossby wave emanate. A similar phenomenon can also occur over the Indian Ocean [59], providing a link between the African and Indian monsoons [60]. Occurring in the boreal summer, these waves meet over West Africa, where they spawn convection and modulate the wind regime, easterly wave activity, and moisture transport [60–62]. Of the two waves, the westward propagating Rossby waves appear to be most important for West Africa [59, 63,64]. The intraseasonal variability peaks in two distinct frequency bands, 10 to 30 days and 30 to 90 days. The former is best developed in the Sahel region. Variance is maximized around 15 days [61, 65], so that it may be part of the quasi-biweekly zonal dipole in convection identified by Mounier et al. [66].

The MJO does influence Sahel rainfall [63, 65], as well as the systems that modulate its variability, such as AEWs, the AEJ, and the low-level monsoon westerlies [64]. However, the MJO's greatest impact is in the latitudes to the south of the Sahel, a result of the near-equatorial track of both the Kelvin and Rossby waves. Intraseasonal variance in the Sahel itself is comparatively small [67], so that the MJO's overall impact on Sahel rainfall (as opposed to the monsoon overall) may be small. There also seems to be little relationship between interannual variability of MJO amplitude and the year-to-year variability on the 30 to 90 day time scale. This has not yet been examined for the shorter 10 to 30 day time scale, which has higher variance in the Sahel.

Several other factors modulate the intraseasonal variability of the West African monsoon. These include two-way interactions between the AEJ and transient systems [68], convection in the "trigger" region near Darfur [69], and intraseasonal variations of the Saharan heat low on the 10 to 25 day time scale [70]. The SHL variations appear to be linked to mid-latitude events [71], such as Mediterranean cold surges [72] and dry air intrusions [73]. Northward bursts of the monsoon also operate on intraseasonal time scales, bringing summer rainfall as far north as the Hoggar [35].

INTERANNUAL VARIABILITY

The interannual variability of rainfall in the Sahel became a topic of great concern as a result of the drought conditions that began in the late 1960s. One of the results of early research on the region was identification of several important—and perhaps unique—features of rainfall variability in the region. These include an extremely large spatial scale of annual and decadal anomalies [7, 74], decadal and interdecadal persistence of anomalies that is especially pronounced in the central Sahel [4, 6, 75], extreme magnitude of variability [76] and a dominant contribution of only two of the rainy season months (August and September) to the interannual/interdecadal variability [6, 77]. This section commences with a description of the variability of annual rainfall over the course of the last two centuries, including the debate about recent trends, and follows with a discussion of these general spatiotemporal characteristics of Sahel rainfall.

Temporal Trends

There has been nearly unanimous agreement about the course of Sahel rainfall from the 1950s to the 1990s (e.g., [5, 78–80]). Indices derived independently from diverse data sets are strongly correlated [81] and indicate that the 1970s and 1980s were markedly drier than the two previous decades. The most intense drought occurred in the early 1980s. Recent work [82, 83] puts these droughts in the context of fluctuations during the last two centuries (Figure 10) and suggests that even more extreme conditions prevailed in most of the first half of the nineteenth century.

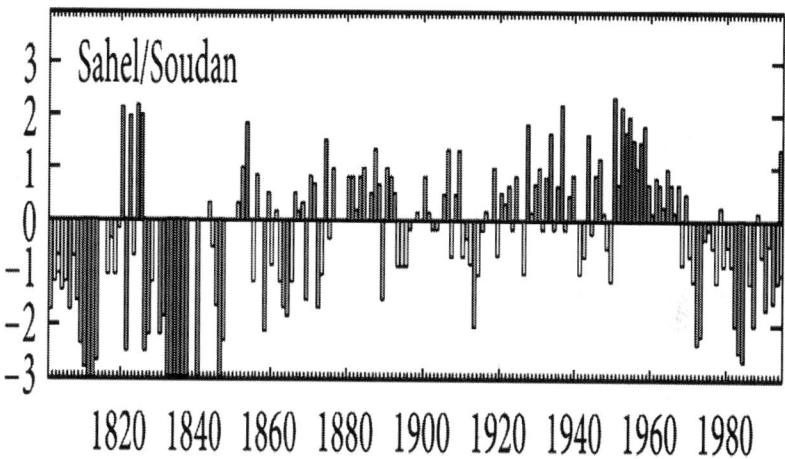

Figure 10: Precipitation variations in the Sahel/Soudan region of West Africa during the nineteenth and twentieth century. Precipitation is indicated as a "wetness" index, which varies from −3 to +3 (from [82]).

Early works on the region's rainfall variability described two prior periods of drought within the twentieth century. The first was in the 1910s and the second was in the 1940s. The roughly 30-year spread between each drought interval unfortunately gave way in the 1970s to speculation that a 30-year cycle existed in the region and that it was robust enough to serve in a predictive capacity (e.g., [84]).

The recurrence of drought in the 1980s showed the fallacy of that idea. It also demonstrated that the "droughts" of the 1910s and 1940s were weak, compared to those of the 1970s and 1980s (Figure 10). In the early 1980s rainfall fell to some 60% of the long-term mean [3]. In contrast, rainfall conditions were very good in the 1950s and early 1960s.

Conditions post-1990 have received less attention, in part because of the difficulty of obtaining gauge data since that time. Nicholson et al. [76] and L'Hôte et al. [81] conclude that the drought continued through the 1990s, with the exception of high rainfall in 1994 and

1999. Ozer et al. [86] suggested that, in fact, the region was entering a more humid period, citing flooding as a point of evidence. L'Hote et al. [79] conclusively showed that the flooding had causes other than high rainfall. Hastenrath and Polzen [87] show a continual increase of rainfall since the 1980s, with several very wet years in the 1990s. Fontaine et al. [88] concluded that deep convection significantly increased since the mid-1990s and also moved northwestward, in conjunction with a similar shift in the Saharan Heat Low.

The most comprehensive analyses of more recent conditions are those of Nicholson [3], Ali and Lebel [2], and Lebel and Ali [8]. These analyses, as well as that of Mahé and Paturel [89] generally indicated conditions of improved rainfall in the late 1990s and a few following years. A greening up in the regime, as shown by vegetation index data, seemed to confirm this (e.g., [90–93]). However, the degree of improvement varied geographically.

Nicholson [3] examined the season as a whole and the month of August for eight sectors of the Sahel. These included four latitudinal strips, each spanning two degrees of latitude, and the strips were divided into east and west sectors (with a division at 15° E). Updating century-long gauge time series with TRMM satellite data for the years 1999 to 2003, she concluded that recovery was greater in the western Sahel (west of 15° E) than in the eastern Sahel and that the recovery was relatively weak in the month of August (Figure 11). In all but the northern-most regions, the rainfall during the period 1998 to 2003 was roughly comparable to that of the 1950s wet period. In the east the patterns in rainfall were more strongly dependent upon latitude. The years 2000 to 2003 were relatively dry in the north and near the long-term mean in the central Sahel, while the years 1994 to 1999 were relatively wet and comparable to the 1950s. In the southern-most latitudes of the eastern Sahel, 1998 to 2003 appear to have experienced record rainfall.

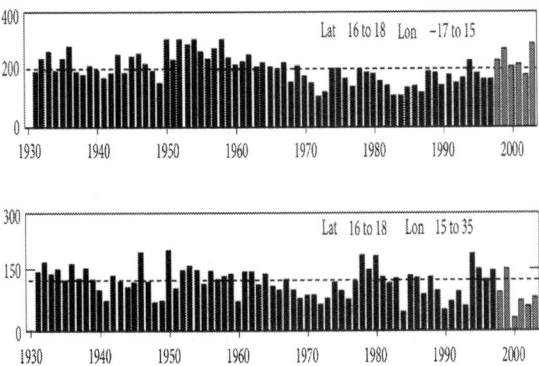

Figure 11: Mean annual rainfall (mm) in the eastern (top) and western (bottom) Sahel in the latitudinal band from 16° N to 18° N (from [3]).

A weakness of that study was that the rainfall conditions of the years 1998 to 2003 were established from satellite data from the Tropical Rainfall Measuring Mission (TRMM), while earlier rainfall conditions were established from gauges. However, a validation of TRMM that compared gauges and satellite estimates for the year 1998 [94] showed zero bias in the TRMM estimates (i.e., the magnitude of TRMM matches that of the gauges) and the root mean square error of the TRMM estimates is 0.7/0.9 mm/day for seasonal/August rainfall.

Lebel and Ali [8] provided further detail, examining conditions through 2007 and considering three longitudinal sectors within the region termed "western Sahel" by Nicholson. Their time series, shown in Figure 12, were more homogeneous, as they were produced solely from gauge data. Some recovery was shown in recent years, but it was confined to the eastern-most regions. The drought continued unabated in the west (longitudes 10° W to 15° W), where mean rainfall for the period 1990–2007 was roughly equal to the mean for the acknowledged drought period 1970–1989. In the central Sahel (0 to 5° E) there was limited recovery. Rainfall for 1990–2007 was only some 10% above the mean for 1970–89 and well below the average for the period 1950–1969. That study also noted the disappearance of the pronounced August peak, consistent with the finding of Nicholson [3] that the recovery in August was very weak. In addition, interannual variability was greater in recent years than in the previous 40 years [2, 87].

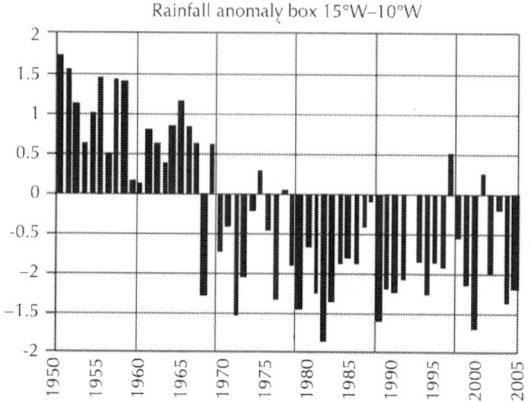

(a)

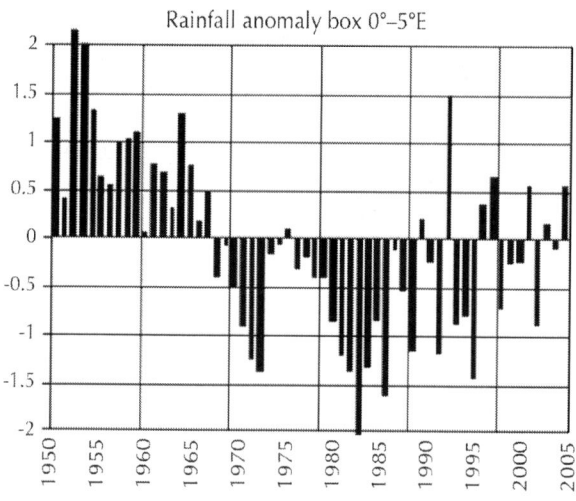

(b)

Figure 12: Rainfall anomalies in the far western and central Sahel, with rainfall expressed in units of standard deviations (from [8]).

Few studies have examined post-2007 years. Fontaine et al. [88] suggest that in 2008 rainfall in the central Sahel was near or just above the long-term mean. A website maintained by the University of Washington (http://jisao.washington.edu/data/sahel/) confirms this for the region as a whole and indicates near normal rainfall in 2009, extremely high rainfall (comparable to the 1950s) in 2010 and abnormally dry conditions again in 2011. Fontaine et al. [88] also showed that the circulation characteristics associated with previous wet conditions in the Sahel have again been occurring since the 1990s. These include a reinforcement of the low-level westerly winds south of the Sahel, a northward shift in the latitude of the African Easterly Jet, and an intensification of the Tropical Easterly Jet [30, 36].

Another controversial point concerning interannual variability is whether, in fact, the long, dry interval has ended [101]. Nicholson [3] confirmed a recovery of wetter conditions in much of the Sahel during the period 1998 to 2003. Ozer et al. [86] also indicated that the Sahel drought ended in the 1990s. On the other hand, L'Hôte et al. [81], examining a Sahel rainfall index that extends from 1896 to 2000, concluded that the drought continued through the 1990s.

The controversy stems in part from changes in the character of rainfall conditions during the last decade or so. The spatial continuity was much weaker than in previous years, as was the year-to year-persistence. Thus, the answer to whether or not the drought ended in the 1990s depends in part on the region, years, and even months evaluated.

Spatiotemporal Characteristics of Rainfall Variability in the Sahel

The change in rainfall conditions starting 1968 was very abrupt and very extreme. For the Sahel latitudes of ~15° N to 20° N mean August rainfall from 1968 to 1997 was roughly 37% lower than the mean for the thirty-year period 1931 to 1960. In the Sahelo-Saharan zone just to the north, the reduction was 55% [76]. There is probably nowhere else in the world where decadal means change so dramatically. There was a general shift of the rainfall isohyets of roughly 1 to 2 degrees of latitude [5, 102]. Several early studies (e.g., [77]) showed the decline in rainfall is associated mainly with conditions in August and September, despite the fact that those months contribute only 50% to 60% of the rainfall in the Sahel.

A related, perhaps unique, characteristic of interannual variability in the Sahel is its decadal to multidecadal persistence [4, 75]. Rainfall was near or above the long-term mean in every year from 1950 to 1967 (Figure11). It dropped abruptly in 1968 and remained below the long-term mean in every year from 1969 to 1997 [3]. This persistence was originally thought to reflect the role of land-surface forcing in perpetuating rainfall anomalies (e.g., [102]). One argument in favor of this interpretation is that the persistence is strongest in the central Sahel, away from the influence of either the Atlantic or Indian Oceans or the Mediterranean [6]. More recent work has shown the issue to be much more complex. Related issues are further discussed in Sections 7 and 8.

The spatial coherence of annual and decadal anomalies is also exceedingly strong. This is illustrated in Figure 13, which shows anomalies in a wet year (1958) and a dry year (1988) [37] and for the decades of the 1950s and the 1980s. In 1955 rainfall was 50% to more than 100% above the long-term mean at the vast majority of stations. In 1983 negative anomalies of that same magnitude prevailed at nearly every station. For the 1950s decade rainfall was well above the long-term mean at stations throughout the Sahel, while the opposite pattern prevailed in the 1980s at virtually every station throughout the Sahel and Guinea coast [102].

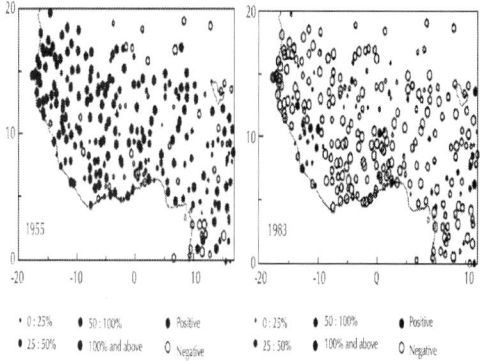

(a)

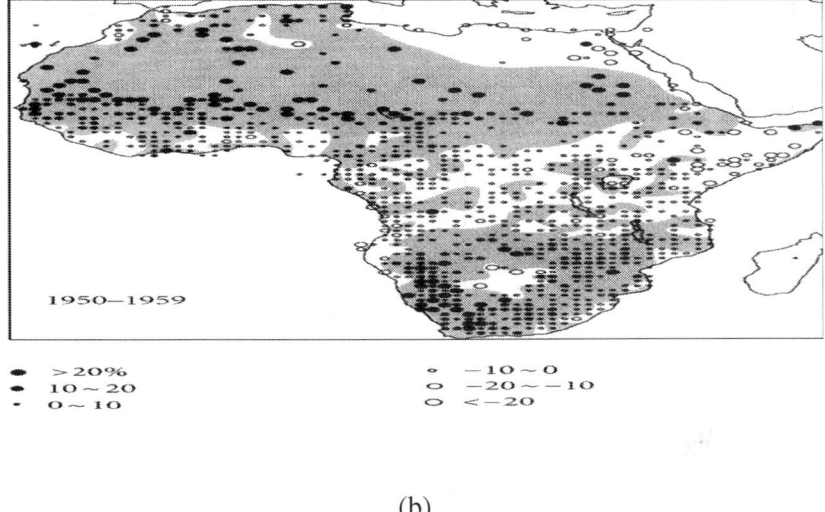

(b)

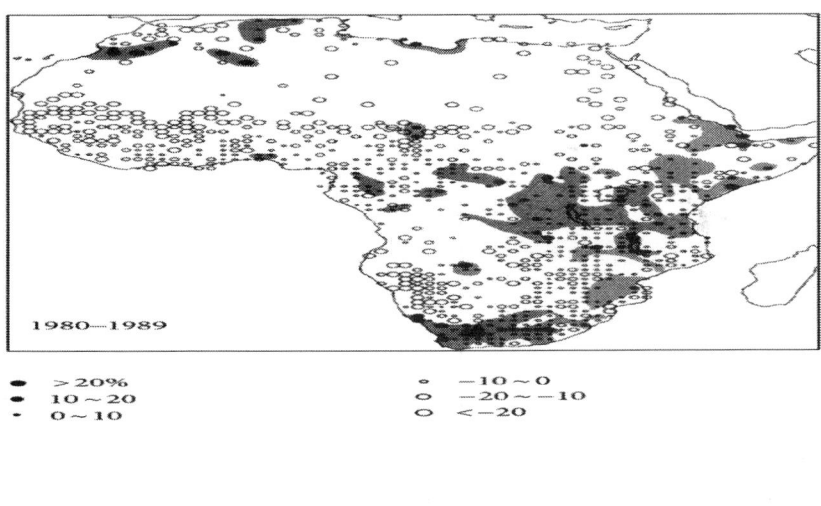

(c)

Figure 13: (a) Rainfall in 1983 and 1955 during the West African rainy season (June through September) (expressed in units of standard departure, based on means for the period 1968–1997) (from [37]). (b) Same but averaged for the decades 1950–59 and 1980–89. In (a) circles indicate rainfall stations; size of the circle indicates magnitude of the anomaly, with open circles indicating negative anomalies and shaded circles indicating positive anomalies (from [76]). In (b) and (c) circles indicate one degree averages.

The pattern of opposition between the Sahel and the more humid Guinea coast to the south, termed the rainfall dipole, is one of two major modes of rainfall variability over West Africa. The other is anomalous conditions of the same sign throughout the region. Nicholson and Grist [103] developed a conceptual model of these modes (see also [36]) that relates the dipole to a latitudinal shift of the tropical rainbelt over West Africa and relates the mode with anomalies of uniform sign to intensification or weakening of the tropical rainbelt. The spatial patterns and latitudinal distribution of rainfall associated with these modes are shown in Figure 14. Notably, the dipole pattern appears to have disappeared in recent decades [104].

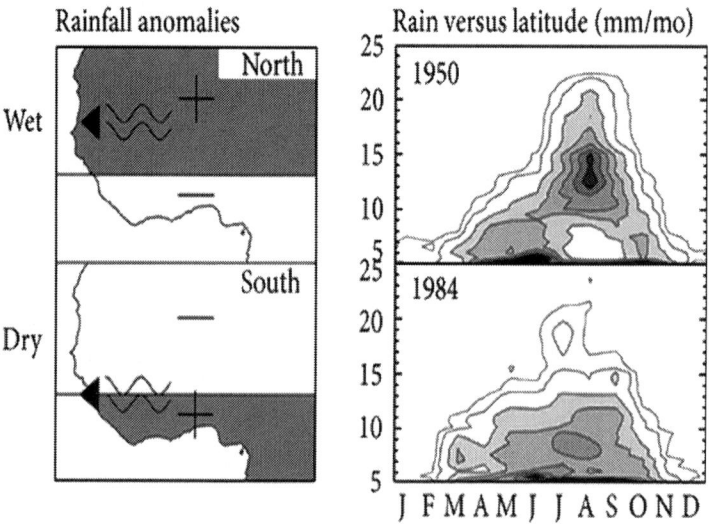

Figure 14: (left) Schematic illustrating the four most common rainfall anomaly patterns over West Africa (from [36]). Light shading indicates below normal rainfall; dark shading indicates above normal rainfall. These are associated with a north or south displacement of the tropical rainbelt or an intensification or weakening of the tropical rainbelt over West Africa. The AEJ position is also indicated. (right) Rainfall as a function of latitude in the months of June through September for 1950 and 1984 (dipole years with a north/south displacement of the rainbelt, respectively and 1955 and 1983, non-dipole years with an intensification/weakening of the rainbelt, respectively).

Although the spatial coherence is strong in both the north-south and east-west directions, notable differences appear between the western, central, and far eastern Sahel. This was particularly noteworthy during the recovery from the long drying trend [3, 8]. The contrasts reflect differences in the dominant circulation systems (e.g., degree of development of the TEJ versus AEJ) [1, 36], the location with respect to the Saharan Heat Low, and the degree of control by the various oceans (see Section 7).

There is some disagreement as to whether or not there is a relationship between the number of rainfall events and year-to-year variability. There is, however, a strong consensus that it is mainly changes in the peak rainy season months that determine the year's character [6, 56, 58, 105]. Bell and Lamb [56] looked at individual convective events and found a relationship between event size/organization and total annual rainfall. They found that the long decline in Sahel rainfall since the 1950s was linked to a decrease in both size and intensity of events, but not to the overall number of events. Nicholson [102] and Lebel et al. [22] similarly found that the annual rainfall in the Sahel is related primarily to the number of occurrences of very intense systems. For example, the difference between a wet August and a dry August at Sahelian stations could be merely one or two events with rainfall on the order of 40 mm to 180 mm per day. In the dry August there were few events with more than 50 mm per day at the stations examined. However, Le Lay and Galle [58] concluded that the long decline in annual rainfall was linked mainly to a decrease in the number of rain events, in agreement with Balme et al. [106]. Frappart et al. [107], examining various locations in Mali, suggested that the role of the number of events varies by region.

ATMOSPHERIC CIRCULATION AND LINKS TO RAINFALL AND ITS VARIABILITY

Some of the earliest studies on interannual variability in the Sahel related changes in rainfall primarily to anomalous latitudinal excursions of the ITCZ [108, 109]. Other studies clearly demonstrated that the link between ITCZ position and Sahel rainfall was tenuous at best (e.g., [7, 110]), but a relationship to the intensity of the ITCZ (as defined by rainfall intensity) was apparent.

A major paradigm shift has occurred, in that it is now well established that interannual variability instead is linked to changes in higher-level circulation features. These include the African Easterly Jet (AEJ), the Tropical Easterly Jet (TEJ), and two low-level westerly jets, the African Westerly Jet (AWJ) over the continent and the West African Westerly Jet (WAWJ) over the Atlantic. The existence of the low-level jets was only recently established. Important circulation features in the region also include a low-level nocturnal jet, the low-level Bodele jet over the Sahara, and the Saharan heat low. Each of these features and its climatic importance, particularly with respect to interannual variability, is described here. Several of them are evident in the zonal mean circulation shown in Figure 15. The spatial configuration of the three major jets is depicted in Figure 6. The TEJ lies in the upper-troposphere around 150 to 200 mb, the AEJ lies in the mid-troposphere at 650 to 700 mb, and the AWJ is in the lower troposphere, with a core at roughly 850 mb.

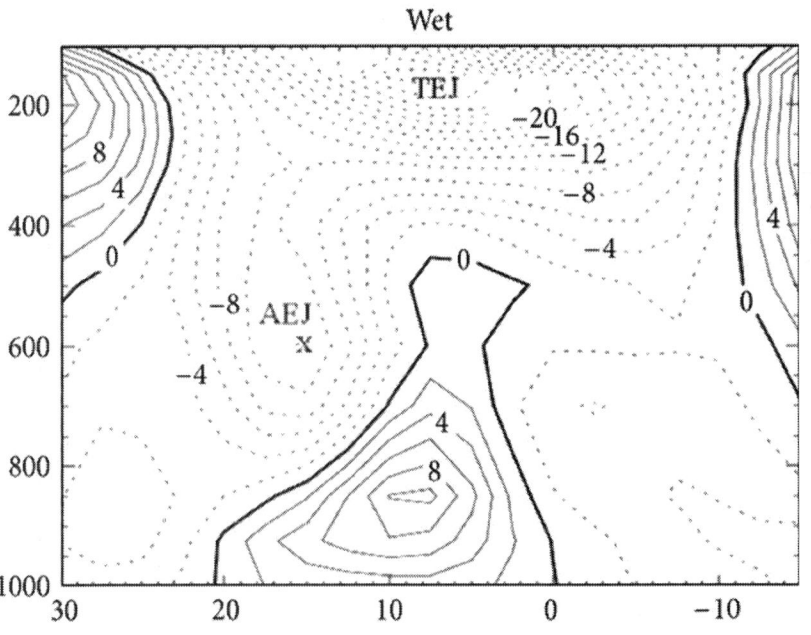

(a)

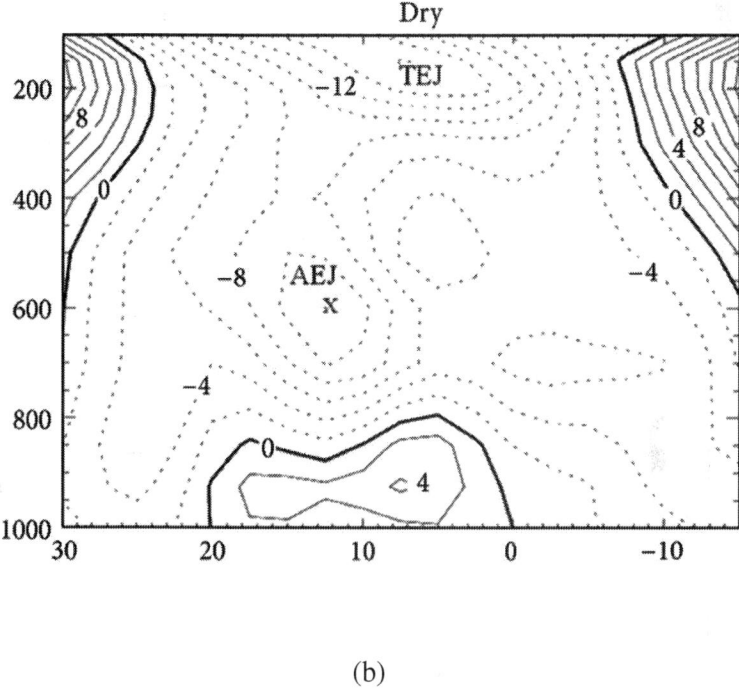

(b)

Figure 15: Vertical cross-sections of zonal wind (m s^{-1}) in wet and dry years. Dashed lines/solid lines correspond to easterly/westerly winds. The tropical easterly jet (TEJ) and the mid-tropospheric African easterly jet (AEJ) is indicated (from [30]).

African Easterly Jet

The African Easterly Jet has received the most attention because of its role in generating African Easterly Waves. It develops in response to the latitudinal temperature gradient between the Sahara and the Atlantic Ocean to the south [1]. Consequently it has a strong east-west orientation. The temperature gradient is particularly strong from June through August, while the AEJ exhibits a maximum intensity in May and June, prior to the onset of the Sahel rainy season. Chen [111] suggests the AEJ owes its existence to the Saharan High, along the periphery of which it lies. This is consistent with the role of the temperature gradient. Moist processes contribute to its intensification and meridional extent Cornforth et al. [112]. Aerosols also influence its development [113].

The AEJ is maintained by both the deep and shallow meridional circulations associated with the monsoon, driven by the local Hadley circulation and the Saharan Heat Low, respectively [31, 114]. The role of the heat low is most important during the pre- and postmonsoon seasons [115]. During the peak monsoon boundary layer processes become the dominant control. Chen [111] suggests an alternate explanation for its maintenance: Coriolis acceleration associated with divergent northerlies in the mid-troposphere over the Chad-Sudan region.

Burpee [116] established that the characteristics of the AEJ meet the criteria for combined barotropic-baroclinic instability. Since the time of his analysis it has thus been assumed that this instability is the major factor in the development of easterly waves over West Africa. More recent studies have suggested that this instability is a necessary but insufficient condition for wave development (e.g., [117]). The AEJ is also unstable with respect to convective heating, as shown by a reversal of the potential vorticity gradient in the vicinity of the jet [117, 118].

The AEJ is usually treated as a single entity, with zonally symmetric structure and a core lying on average over the western Sahel. Thus, many studies of the jet have assumed a uniform basic state. Hall et al. [119] demonstrate that, instead, the AEJ has a zonally varying structure. Moreover, this jet appears to have two independent cores that merge on occasion [120]. The eastern core is somewhat weaker and several degrees further equatorward.

Nicholson and Grist [1] derived the mean characteristics for both the eastern and western sectors of the AEJ. The jet is strongest in May and June, when its mean core speed is on the order of $12\,\mathrm{m\,s^{-1}}$ in the western sector (10° W to 10° East) and $10\,\mathrm{m\,s^{-1}}$ in the eastern sector (10° East to 30° East). Speeds decrease by some $2\,\mathrm{m\,s^{-1}}$ during the Sahelian phase of the monsoon in the boreal summer. The mean monthly speed of either core can attain $16\,\mathrm{m\,s^{-1}}$ during the boreal summer. The intensity of both cores changes markedly from year to year, with the western core being more prominent in wet years and the eastern core being more prominent in dry years.

The latitude of the western core is relatively stable from year to year, on average about 16° North during the boreal summer [120]. The latitude of the eastern core varies markedly from year to year and is inversely correlated with core intensity. Its mean latitude over the eastern Sahel is 13° N during the boreal summer, but ranges from 20° N in some wet years to 10° N in some dry years. In the boreal winter both cores shift equatorward to about 2° N.

Intraseasonal shifts are also apparent in the latitude of the AEJ's core. This occurs in response to the passage of Mesoscale Convective Systems, which cause a north-south split in the jet [121]. An "effective Coriolis force" associated with southerly flow at jet-level contributes to the formation of the secondary jet while the northward flow south of the MCS weakens the original jet. This is a meridional circulation associated with the MCS. The original jet is displaced southward and a secondary jet forms to its north, consistent with the split jet observed in JET2000 [13].

Tropical Easterly Jet

Much less work has been done on the Tropical Easterly Jet, with most studies focusing on the TEJ in the Asian sector, where its core lies (e.g., [122–124]). The TEJ develops as a response to the intense north-south temperature gradient between the Himalayan plateau and the Indian Ocean. It is energetically maintained via tropical divergent circulations associated with the east-west Walker circulation and north-south Hadley circulation. The TEJ's mean speed over the eastern Sahel during the boreal summer is roughly 18 m s^{-1}. It has frequently been described as a boreal summer feature, but this jet is also very strong in January through March, when it is located in in the Southern Hemisphere with a core around 5 to 10° S [1]. In contrast to the AEJ, the TEJ over West Africa exhibits little latitudinal change from year to year but its speed and east-west extent vary greatly [30, 36, 103, 125].

The role played by the TEJ in interannual variability has been well documented. A stronger/weaker TEJ has been linked to wetter/ drier conditions in the Sahel (e.g., [30, 36, 126]), western equatorial Africa [82, 120], Ethiopia [127], and India [124] and to a decreased/ increased number of cyclones in the Bay of Bengal [128]. Its link to rainfall in equatorial Africa is apparent in both the boreal spring and boreal autumn. It also has been implicated as a factor in the mean climate of West Africa [129–131].

The relationship of TEJ intensity to interannual variability is particularly strong in the Sahel. In some wet years its mean speed over the Sahel in August approaches $30\,\mathrm{m\,s^{-1}}$ (Figure 15) and the jet is evident across the east-west extent of Africa. In dry years, its maximum speed over the Sahel can be less than $10\,\mathrm{m\,s^{-1}}$ and it is evident only in the eastern Sahel. The TEJ's speed and spatial extent have diminished between the 1950s and the 1990s, commensurate with drier conditions in the Sahel [103, 125, 128].

Unfortunately, the reasons for its change in intensity are not fully understood. Chen and van Loon [132] noted that the TEJ tends to be weaker during warm phases of the Pacific ENSO. Dezfuli and Nicholson [82,100] noted that its intensity appears to change in relationship to the intensity of the extratropical Southern Hemisphere westerlies during the May-to-June and October-to-December seasons. That seems to be the case during the Sahel rainy season of the boreal summer as well [30]. The stronger jet is also commensurate with cooler upper-troposphere tropical temperatures and latitudinal temperature gradients [36].

The link between the TEJ and rainfall appears to be primarily upper-level divergence associated with the jet core. The divergence results from strong meridional components associated with the TEJ over Africa [1]. This pattern is particularly pronounced in wet years in the Sahel, and this appears to play a role in its impact on rainfall [37]. This is further consistent with the observation that a stronger TEJ is linked not only to more rainfall in the Sahel, but also a more intense rainbelt (e.g., [36, 37]).

An additional mechanism may relate to the dynamic instability of the TEJ [133, 134]. Both barotropic and combined barotropic-baroclinic instability have been demonstrated. This permits the development of waves on the TEJ. Such waves have been observed over West Africa [135] and their characteristics match those predicted by a numerical simulation of African wave activity [136]. Such waves may play a role in the development of extreme precipitation events in the western Sahel.

Low Level Jets

Bodele Jet

The third easterly jet in the circulation regime over Sahelian Africa is the Bodélé Low-Level Jet. It was first documented by Washington and Todd [137] and was shown to be a factor in the tremendous amounts of dust generated in the Bodélé Depression to the north of Lake Chad. The jet is present in all months except August, but is strongest in January and weakest in July. Its mean speed in the core is 8 m s^{-1}. Its easterly flow also has a weak northerly component. The Bodélé Low Level Jet has a strong diurnal cycle, with peak winds occurring in the evening and minimum speeds during the day [138].

The jet's core lies near 18° N and 20° E at 925 mb. This location is the exit region of the gap between the Tibesti and Ennedi massifs. The jet is absent further west where the topography becomes relatively flat. Hence orography effects play a role in its development [139, 140].

African Westerly Jet

The low-level African Westerly Jet (AWJ) was first described by Grist and Nicholson [30]. It is not apparent in all years. Rather, its development appears to be limited to wet years in the Sahel. In these years it is on the order of 10 m s^{-1} during the late boreal summer (July to September) and it extends well into the mid-troposphere. In dry years it all but disappears in the monthly mean and the wind shifts to easterly above the 850 mb level. The westerly maximum then lies in the monsoon layer at about 925 mb and westerly speeds are on the order of 2 to 4 m s^{-1}.

The AWJ is not simply part of the southwest monsoon flow. There is a marked directional discontinuity, with the disappearance of the southerly component [1]. Moreover, when the AWJ is apparent, its core is well above the monsoon layer. During August, the height of the Sahel rainy season, the core speed of the AWJ is well correlated with the surface pressure gradient between the latitude of 20° S and 20° N (r = .84 over 58 years). However, the jet is not a geostrophic response to this gradient. The region is too close to the equator for geostrophic balance and the jet develops in a narrow belt in the sector where the cross-equatorial pressure gradient is weakest. Its origin appears to be linked to inertial instability that develops in response to this pressure gradient [95, 141]. The inertial instability enhances both the jet and rainfall. The correlation during August between the AWJ speed and Sahel rainfall during the same 58 years is .75.

West African Westerly Jet

A low-level westerly jet over the Atlantic was first described by Grodsky et al. [142]. It appears as a near-surface wind maximum at over the equatorial Atlantic and is evident from May through September. This jet, termed the West African Westerly Jet (WAWJ) by Pu and Cook [143], is best developed around 10° N and lies in the region where the trade winds converge. Not to be confused with the AWJ jet over the continent described by [144], this marine jet appears to be the surface manifestation of a mid-tropospheric westerly wind maximum. Its speed can reach 10 to 15 m s^{-1}.

The importance of the WAWJ may lie in the moisture it transports from the Atlantic to the continent in the zone from 8 degrees to 11 degrees north. Although the transport does not extend into the Sahel, there is a strong correlation between the speed of this jet and western Sahel rainfall [142]. The moisture transport by this jet has much greater decadal-scale variability than that associated with the southwest monsoon and it may bring moisture into the region in years when the monsoon is weak [145]. The West African Westerly Jet is also important for stabilizing the regional vorticity balance by introducing strong relative vorticity gradients.

Nocturnal Jets in the Monsoon

Examination of the boundary layer at several locations within the West African monsoon showed the existence of a nocturnal low-level jet (NNLJ) [29]. It lies generally within the layer 200 to 400 m above the surface [146, 147]. Studies at Niamey, Niger (c. 14° N) and at Nangatchori, Benin (c. 10° N) showed that this feature tends to occur throughout the year. The nocturnal jet reaches a maximum around the onset of the monsoon and disappears late in the season when the rainbelt commences its southward migration [148].

The development of the jet is linked to the diurnal variation of the boundary layer [29]. This cycle is maximized in the northern part of the monsoon layer, where the meridional pressure gradient is strongest. Typical of arid and semi-arid regions, the West African boundary layer is associated with daytime turbulence and comparatively laminar flow at night. Mean daytime wind speeds are relatively low. An inversion layer develops after sunset, following by the development of the jet. Maximum speeds are commonly on the order of 8 to $10\,m\,s^{-1}$, but can exceed $15\,m\,s^{-1}$ [147, 149]. The direction of the NLLJ depends on which side of the Intertropical Discontinuity it occurs, being generally southwesterly when imbedded with the monsoon flow and northeasterly when imbedded within the Harmattan [29, 147].

The NLLJ transports moisture into the monsoon, especially around its onset [148]. This helps to sustain deep convection [150]. It also appears to be responsible for the stratus cloud decks that often form in the Soudanian and Guinean climate zones of West Africa (i.e., south of the Sahel to the Guinea coast) [151].

Saharan Heat Low

The presence of an intensive heat low in the western Sahara has long been established. Generally assumed to be a passive component of the monsoon system, the Saharan Heat Low (SHL) has been shown by recent research to be a key element in the monsoon system and its variability. A causal link between the strength of the heat low and Sahel rainfall exists on intaseasonal [152], interannual and decadal time scales [153].

The heat low, also called the West African heat low, is a region of exceedingly high surface temperatures and low surface pressure [154]. It is a shallow system, a thermal depression generally below 700 mb [26]. While present throughout the boreal summer, it undergoes synoptic and intraseasonal changes in intensity.

The Saharan Heat Low shows a pronounced seasonal migration. Starting from a location south of the Darfur mountains in the boreal winter (November to March), it migrates northwestward to a location in the western Sahara. It maintains approximately the same location, between the Hoggar and the Atlas mountains, between late June and

September [154]. It generally reaches this position about 5 days prior to the onset of the West African monsoon.

Chauvin et al. [71] suggest that the SHL is a bridge between the midlatitudes and the West African monsoon. They identify what they term a west phase and an east phase. In the west phase, maximum temperature is over the coast of Morocco-Mauritania and a minimum appears between Libya and Sicily. The east phase has the opposite temperature structure. The west phase is preceded by large-scale intraseasonal fluctuations in the mid-latitudes, such as Rossby waves. It is concomitant with enhanced convection over the Darfur region, an area shown to trigger African Easterly Waves (e.g., [155]).

The intraseasonal fluctuations in the intensity of the Saharan Heat Low are concentrated in two bands, 3 to 10 day and 10 to 30 day [70, 152]. Variance in the higher frequency band peaks at roughly 5 days [156]. This suggests a link to wave development and convection, but the link is not clear. Lavaysse et al. [152] identified strong and weak phases of the heat low and evaluated circulation, temperature and convection within them. They showed increased/decreased convection in the central/western Sahel in the strong phase, in response to an increase/decrease in the low-level cyclonic circulation around the heat low. Hence, changes in the intensity of the SHL affect the intensity of the monsoon circulation further south [29].

Lavaysse et al. [26] noted widespread convective activity in the Sahel during a two-week period in September 2006 in which the intensity of the Saharan Heat Low was abnormally weak. This apparent contradiction appears to have resulted from the interaction of the SHL with a mid-latitude depression that brought a cold surge into Libya at 700 mb, weakening the low but increasing convection over the Sahel. Hence there are mid-latitude influences on the SHL.

The Saharan Heat Low is also influenced by dust, rainfall, and the inflow of relative cold and statically stable air from the Atlantic. The effects of the dust are two-fold [157]. A direct effect is related to radiative heating, which increases the thickness of the heat low. The second effect is related to the impact of the dust on the AEJ and African easterly waves. The dust intensifies both and through this enhances the intraseasonal variability of the SHL. The inflow from the Atlantic, mainly in the form of a mesoscale sea breeze and associated front, creates a baroclinic zone to the west of the SHL by bringing in cool and moist air [158]. Cuesta et al. [35] noted a conspicuous weakening of the low following rainfall events.

Saharan Air Layer

The daytime Sahelo-Saharan boundary layer has a split structure with a well-mixed, near-surface convective layer lying beneath a residual layer with more laminar dynamics [159]. This residual layer, termed the Saharan Air Layer or SAL, is a deep layer of hot and dry air characterized by a red haze. The haze is due to mineral dust eroded from the Sahel [160] and is associated with the northeast Harmattan but easterly waves add to the dust concentration in the haze. The east-west extent of the SAL is typically 2000–3000 km, but it can extend some 5000 km, covering an area about the size of the contiguous United States [161]. Where the SAL meets the eastern Atlantic it resides between roughly 900 mb/1800 m and 500 mb/5500 m [85]. It extends even higher over the continent, where its depth decreases from north to south [162].

The SAL is well mixed with relatively uniform conditions of high heat content and low moisture content. A typical temperature in the layer is 44°C. A typical mixing ratio would be 2 g/kg [163]. When the SAL is not present the atmosphere is markedly cooler and more humid. In its absence the relative humidity is some 25 to 45% lower and the mixing ratio some 2.5 to 5.5 g/kg lower [161]. The SAL is bound by subsidence inversions at its top and bottom [85] (Figure 16). These inversions help maintain the integrity of the layer. The dust that defines the layer also helps to maintain its thermal structure [164]. At the top of the SAL lies a region of high relative humidity where altocumulus and stratocumulus layers are often observed [162].

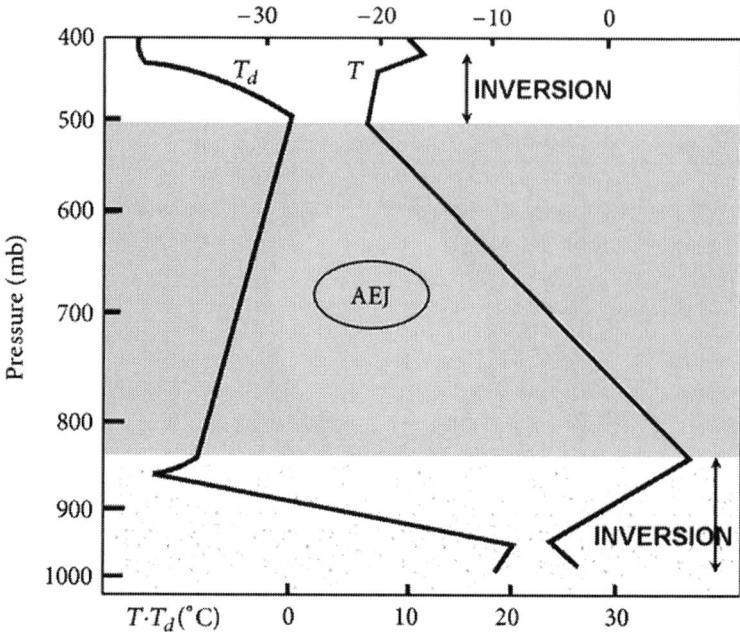

Figure 16: Conceptual model of the Saharan air layer (SAL) over West Africa (from [85]). The Saharan air layer (dark shading, indicating intense dust layer) is trapped between two inversions. Dust is evident below the inversion but in lower concentrations.

LiDAR observations have shown that there are two dust plumes within the SAL. One originates over the northern Sahara and the other from the Bodélé region of Chad [165]. The Bodélé Jet plays a major role in the development of the latter plume [137].

The SAL has significant impact on meteorological processes over the Sahel by way of its radiative effect, its thermal character, and its dryness. It impacts the African Easterly Jet, African Easterly Wave Development [13] and probably the development of tropical cyclones and hurricanes over the Atlantic [161]. The impact of the SAL on the AEJ is a result of the low static stability of the air mass. This creates a negative PV anomaly and anticyclonic relative vorticity on the poleward side of the AEJ. The low is a key part of the PV-sign reversal that characterizes the instability of the AEJ [13, 118]. The north-south temperature gradient within the layer is relatively small aloft and thus creates a "cap" on the AEJ's core, thereby determining its vertical extent [162].

Links of Major Circulation Features to Rainfall and Its Variability

The major circulation features associated with the variability of Sahel rainfall on interannual and decadal time scales are the TEJ, the AEJ, the AWJ, and the Saharan heat low. Nicholson has described the influence of the three jets in several publications (e.g., [20, 30, 36, 37]). These works did not examine the Saharan heat low directly. However the results of Nicholson and Webster [95] clearly imply its role. That study is also consistent with the concept of a mid-season "monsoon jump", when cross-equatorial pressure gradients surpass a critical threshold needed to establish inertial instability.

As described in Section 3.2, the interannual variability of rainfall takes primarily two forms, a latitudinal displacement or a change of intensity of the tropical rainbelt over West Africa. In the former case, the well-known dipole ensues, with rainfall of the opposite sign over the Sahel and Guinea Coast regions. The "node" of the dipole is at roughly 10° N. In the latter case rainfall anomalies are of the same sign in both regions. The commonality in both cases is the intensity of the Tropical Easterly Jet. The core speed is anomalously high when annual rainfall is above average in the Sahel, anomalously low during dry years in the region. The rainbelt over West Africa lies between the cores of the African Easterly Jet and the Tropical Easterly Jet, a region of strong vertical motion and divergence aloft. An intensification of the rainbelt/vertical motion requires a strong displacement of the two cores with respect to each other and strong vertical shear.

The occurrence of the dipole is determined mainly by the location of the AEJ core, which in turn is modulated by the intensity of the low-level African Westerly Jet over the Guinea Coast region. In the dipole case with wet conditions in the Sahel, the AEJ is displaced northward. In the dipole case with dry conditions in the Sahel it is displaced southward of its mean position.

Nicholson [20] suggested that the wet and dry cases in the Sahel represent two independent dynamic modes, with the switch between modes being associated with inertial instability. The main points of evidence supporting this are the bimodal frequency distributions of annual rainfall and dynamics variables related to it and the existence of critical thresholds (Figure 17) that separate the dry and wet years. These thresholds appear to be a cross-equatorial pressure gradient of 10^3 mb/km and an AWJ speed of 7 m s^{-1} [95].

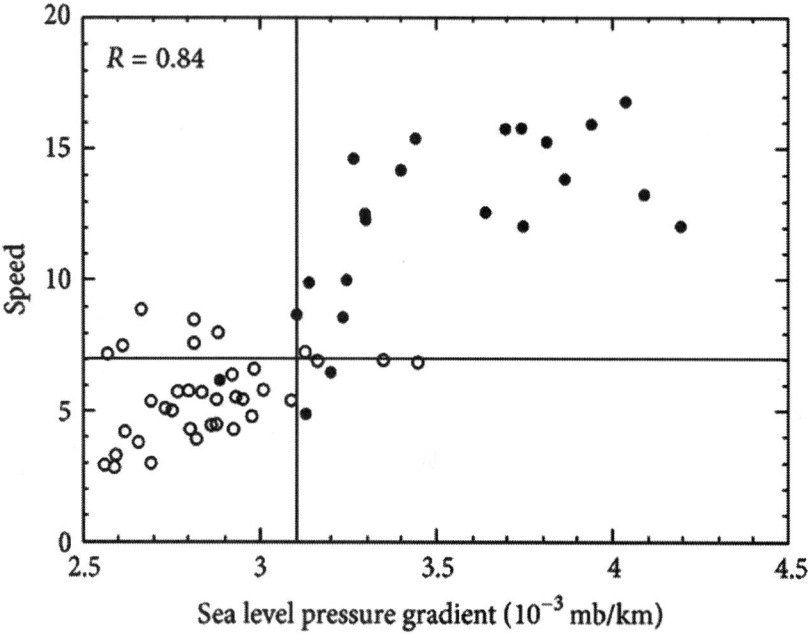

- 1948 to 1969
- 1970 to 2004

(a)

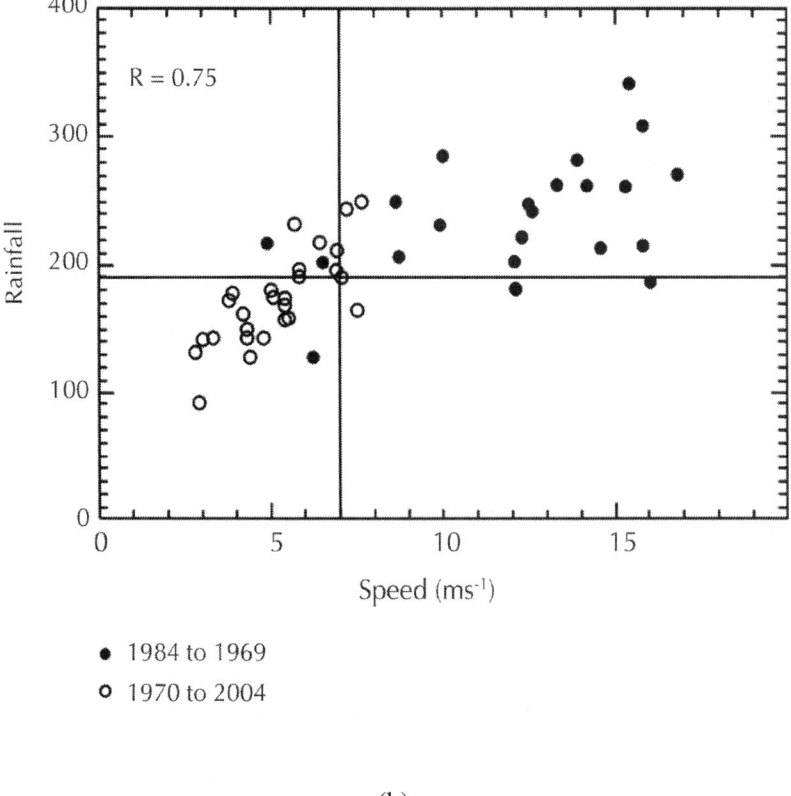

● 1984 to 1969
○ 1970 to 2004

(b)

Figure 17: (a) Speed of the westerlies (m s⁻¹) at 850 hPa versus surface pressure gradient for August of the years 1948–2004 (from [95]). Data are averaged for 5° W to 5° E and the pressure gradient is calculated between 20° N and 20° S. The open circles represent the years 1948–1969 and the solid circles represent the years 1970–2004. (b) Annual rainfall in the Sahel (averaged within the sector 10–18° N and from the Atlantic coast to 30° E) versus speed of the westerlies at 850 hPa (from [20]).

Grist and Nicholson [30] contrast the conditions of the wet and dry mode. The main contrasts are in the intensity of the TEJ and the AWJ and the latitude of the AEJ. However, contrasts are also evident in the structure of the moist layer, the intensity and character of wave activity, and the seasonal evolution of the low-level temperature gradient.

PRECIPITATION AND CONVECTION

Precipitation over the Sahel is related primarily to three types of systems: mesoscale convective complexes, local convective activity, and various synoptic scale systems that represent interactions between the tropics and extra-tropics. The mesoscale systems produce most of the rainfall. Isolated, local, daytime convective activity produces only a small share of the Sahel's rainfall. Although favored by local, near surface conditions of high sensible heat flux, high humidity, and steep lapse rate, the initiation of these local systems is favored by mesoscale ascent [166]. The hybrid tropical-extratropical systems tend to produce precipitation during the Sahel's dry season or in the pre- or postmonsoon seasons.

A typical MCS is shown in Figure 18 and compared with other systems. An MCS is essentially a large, continuous area of deep cloud (at least 2,000 km²) in which one or more areas of convective precipitation are imbedded. The average size of MCSs is 10,000 km². The presence of ice in the upper cloud layers in used by some (e.g., [23]) to distinguish MCSs from other convective systems. Over the Sahel MCSs account for up to 90% of the precipitation and 50% of the rainfall in the tropics, although they comprise only 2% of all rain-bearing features in the tropics [167]. The typical MCS is topped by a large anvil of stratiform cloud (Figures18 and 19). Despite the intense convection associated with these systems, stratiform precipitation accounts for 73% of the rain area and contributes roughly 40% of the total rainfall for the tropics as a whole [168]. Over the Sahel the contribution to total rainfall ranges from less than 20% in the south to more than 80% in the north [169].

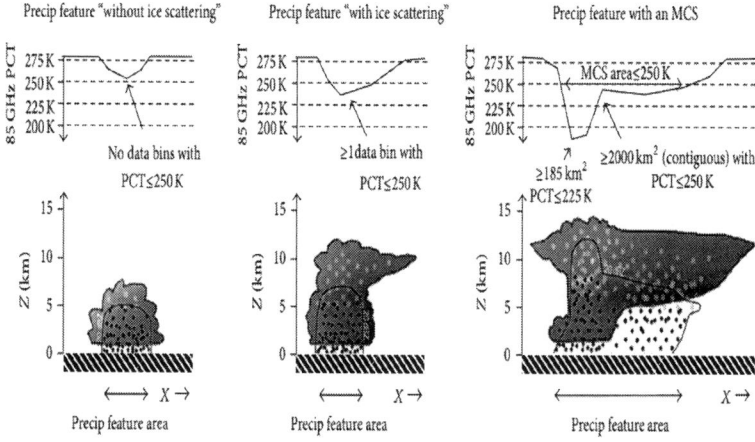

Figure 18: Schematic of typical tropical precipitation features, including mesoscale convective complexes (from [96]).

Figure 19: Photo of anvil associated with an MCS over West Africa (from [57]).

MCSs are in a constant state of evolution. These systems produce intense convective rainfall mainly during the afternoon, with most convective events lasting 3 hours or less over land [170]. As the system evolves in the later hours of the day and into the night, the cloud anvil that tops the system spreads and produces a large area of stratiform cloud [23]. Thus, the stratiform rain typically occurs at night and for a longer period of time, but the rain rate is roughly one fourth the rain rate associated with convective clouds [168].

The mesoscale convective complexes are of two types, rapidly moving squalls and slower moving non-squall systems [168]. During the months of July through September of 1986 and 1987 153 squalls were observed over West Africa [171]. These did not produce much rainfall, only 7 mm per event, on average. A study at Niamey, Niger (13.5° N) during the AMMA special observing period of 2006 found that MCSs with squalls accounted for about 90% of the rain there, though the squalls were present in only 17% of rain events [172]. The squalls occurred along two AEW tracks at around 8 to 16° N and 2 to 6° N. More stratiform precipitation occurred with the squalls moving along the northern track.

Fink and Reiner [173] tracked 81 AEWs and 344 squall lines over West Africa during the boreal summers of 1998 and 1999. They found 42% of the squalls to be wave-forced, a proportion increasing westward from 20% in the eastern Sahel to 68% at the West African coast. Squall occurrence peaks around midnight and has a minimum around noon [171]. Most nocturnal squalls in the Sahel are wave-forced. Wave-forced squalls are particularly frequent in the Sahel and during the peak monsoon season (July to September), when rainfall peaks in the Sahel. Favorable locations of Sahelian squalls are west of the AEW trough and in the region of southerly flow east of the trough. The AEWs appear to be most important in squall initiation, since the AEW-forced and non-forced squalls show little contrast in characteristics [173].

The major issues concerning convection in Sahelian West Africa are its trigger, its association with African Easterly Waves and its association with the African Easterly Jet. Convective episodes tend to arise in the lee of high terrain, consistent with thermal forcing from elevated heat sources [174]. Propagation is generally associated with moderate low- to mid-tropospheric shear, which varies with the latitudinal migration of the AEJ migration and with the phase of the West African monsoon. MCSs are usually, but not always, associated with AEWs. However the location within the wave depends on several factors, such as longitude, latitude, and period of the wave (see Section 6).

The location with respect to the AEJ was evaluated by Mohr and Thorncroft [175]. They separately considered weak and intense convective systems over the Sahel, defined respectively as the lowest and highest deciles in the intensity distribution. The peak location of the intense systems migrated throughout the rainy season, following the seasonal migration of the AEJ. The weak systems tended to reside around 10° N, remaining there from May to August, but shifting eastward as the season progressed. The majority of both weak and intense systems occurred within five degrees of high terrain. Other factors favoring the development of intense convective systems include high potential temperature and high shear in the vicinity of the AEJ. Intense systems with unlikely to occur in wave troughs or in cases of high atmospheric dust-loading in the Sahel [176].

Although most of the rainfall in the Sahel is associated with the West African monsoon and occurs in the peak of the boreal summer, unseasonal rains can occur in the transition seasons and even in the heart of the dry season. These most often affect the western Sahel and can bring as much as 25 mm to the Sahel in the middle of the dry season. A case in point is the "heug" rains of Mauritania. These can persist for days on end [97].

Most of this unseasonal rainfall is associated with systems that develop as a result of tropical-extratropical interactions. The two systems most commonly described are the "tropical plume" [177] and the Soudano-Saharan depression [97, 98]. The common elements of these systems (Figure 20) are a diagonal trough emanating from the mid-latitude westerlies at upper levels and a surface tropical disturbance, often an AEW. Several cases of tropical-extratropical systems over North Africa are discussed in an extensive series of papers by Knippertz (e.g., [177–182], and by Geb [183]. Many are similar to the classical "tropical plumes", but may differ with respect to synoptic scale features or the continuity of the associated cloud bands. Schepanski and Knippertz [181] examined several systems that resembled the classic "Soudano-Sahelian depression" and concluded these cannot generally be traced to surface-level depressions and call for a revision of the classic picture of these systems. Further examination is warranted because such systems may have played a much greater role in the Sahel in past centuries, contributing to some of the historical wetter episodes experienced by the region [97].

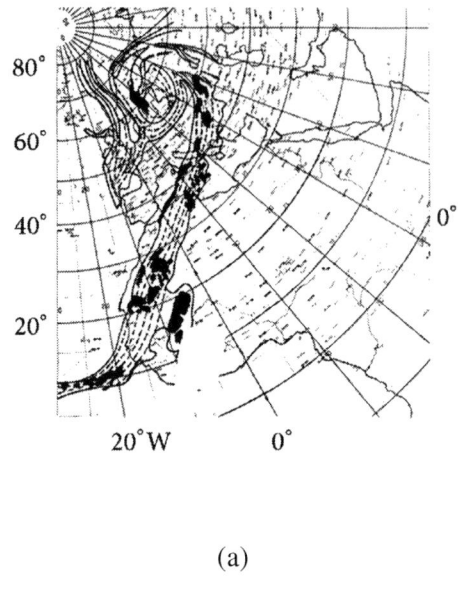

(a)

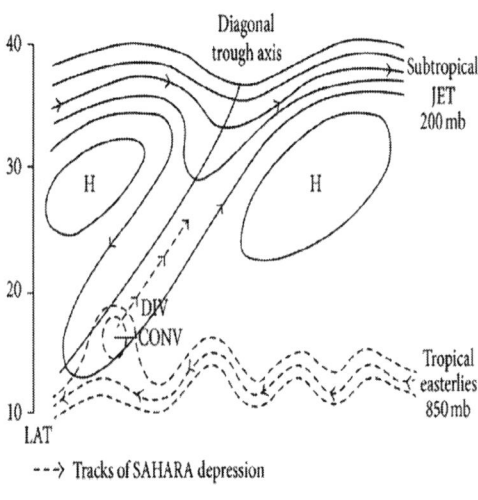

(b)

Figure 20: Development of Soudano-Saharan depressions and diagonal cloud bands over West Africa (from [97, 98]). Diagram on the left depicts a satellite view of a diagonal cloud band and areas of heavy rain within it (shaded areas) in September 1969, a period of tremendous rainfall and flooding in parts of North Africa. Diagram on the right shows the typical circulation pattern that leads to its development.

WAVE ACTIVITY AND LINKS TO CONVECTION

In the mid-1970s it became apparent that a disturbance system termed the African Easterly Wave (AEW) is an extremely important component of the West African monsoon. Classic ideas concerning AEWs include their development from the African Easterly Jet via combined barotropic-baroclinic instability and their role in organizing and/or promoting convection. During the last decade the paradigm has shifted, with many studies emphasizing the mutual interaction between waves and convection, the role of convection as a wave trigger, and the role of additional types of waves on convection over West Africa.

The AEWs are disturbances in the zonal wind field that typically have wavelengths on the order of 2000 to 5000 km [68]. They propagate along two east-west tracks [184] and occur in two distinct frequency bands of 3 to 5 days and 6 to 9 days [185]. Along the northerly track (~18° N to 20° N), poleward of the AEJ core, the waves are at low levels. Waves along that track are associated with a low-level potential temperature gradient between the dry, hot Sahara and the cooler and moist monsoon air to the south [186]. They are seldom associated with precipitation [187] and are termed dry disturbances by some authors (e.g., [188, 189]). Waves on the southern track, at ~9° N to 11° N and equatorward of the AEJ core, are related to a sign reversal in the meridional potential vorticity gradient. They are usually, but not always, associated with convection and precipitation [186, 190]. The flow in the two tracks might not be independent; rather they might represent a single AEW [191].

Simulating wet and dry years with a numerical model, Nicholson et al. [135] predicted that AEWs can trigger easterly wave development on the Tropical Easterly Jet in the upper troposphere. The predicted waves are of planetary scale with a period of 5 to 6 days and they develop to the northwest of the mid-level AEW. Observations showed that such a wave developed in July of 1950, a very wet year (Nicholson et al. [136], with characteristics similar to the predicted TEJ waves.

Potential vorticity theory suggests that the waves on the TEJ develop via interactions between the surface and the TEJ. This is consistent with role of the surface potential temperature gradient in wave development north of the jet and the strong development of TEJ waves in wet years [186], when that gradient is anomalously strong.

A relatively recent shift in our understanding of AEW generation is the hypothesis that the waves are initiated via convection, rather than purely dynamic instabilities (e.g., [155]). Thorncroft et al. [69] formulated this in the form of the "triggering hypothesis", which suggests that AEWs are the result of a finite thermal disturbance that can grow in the absence of an unstable basic state. Convection particularly that over the highlands of Darfur in the eastern Sahel, is suggested as a potential mechanism [69, 192]. The importance of convection is supported by several studies. Cornforth et al. [112], for example, show that moist processes resulted in much more rapid wave growth. Their simulations also showed that moist AEWs were preceded by an increase in mean rainfall. The idea that barotropic-baroclinic instability is not a prerequisite for wave development has been supported by some modeling studies (e.g., [117, 118]).

AEWS are clearly convectively coupled (e.g., [13]). Nevertheless, their structure and development can be largely understood in terms of dry dynamics [68, 119]. Nicholson et al. [136] for example used a dry model to simulate wave development and produced realistic waves for dry years and wet years that matched observed characteristics in those years. Once a wave precursor is triggered, the state of the AEJ still plays a critical role in determining whether or not a wave develops [193]. This suggests that dynamic instabilities are still a prerequisite for AEW development.

Other wave activity over Africa is triggered in the Pacific, generally in association with the Madden-Julian Oscillation (MJO). Convection over the Indian Ocean or the Pacific warm pool in the boreal summer sends Kelvin waves eastward and Rossby waves westward. They meet over Africa, where they modulate the mid-tropospheric temperature structure in ways that favor deep convection [194]. The westerly flow of the Kelvin wave increases boundary layer monsoon flow and moisture supply. The westerly anomalies also increase cyclonic shear on the equatorward flank of the AEJ, enhancing wave development and transient convective activity. MJO-induced waves often trigger convection over the Darfur region [195], where AEWs frequently originate [69]. The Rossby waves appear to be the more important part of the MJO-induced response over Africa, especially on time scales of 25 to 90 days [59, 60, 63].

The Kelvin waves have a wavelength of some 8,000 km and a period of 6 to 7 days. Mounier et al. [196] suggest their impact on rainfall and convection is of about the same magnitude as that of AEWs. Kelvin waves are more active in dry years than wet years because they are favored by the same Pacific SST patterns that are linked to Sahel drought [197]. In dry years they probably contribute more to convection that do the AEWs.

Grist [198] describes the seasonal cycle of AEWs over the Sahel. He shows peak development during July through September, when horizontal shear and barotropic instability also peak. Longer periods waves (6.25 to 7.5 days) provide a greater contribution in the latter half of the season. Some disturbances are apparent at low levels, but generally only the shorter period waves (3.75 to 5 days), which have a maximum variance in July. This is consistent with the results of Cornforth et al. [112], who show that the surface signal of AEWs becomes weak during the peak monsoon season. They attribute that to the interaction between convection and the waves.

The relationship between AEWs and convection is complex. Originally it was thought that they principally organize convection. More recent work has shown feedback relationships in which convection can modulate waves, which in turn modulate convection. Early studies suggested that the mesoscale convection systems (MCSs) that bear most of West Africa's precipitation are often imbedded in the northerly flow ahead of the wave trough (see [194]). More recent work has shown the relationship between AEWs and convection to be much more complex.

In regions of deep convection the 2 to 6 day waves produce some 25% to 35% of the intraseasonal variance in convection [155]. The waves also appear to modulate annual precipitation amounts [136, 144]. Taleb and Druyan [199], in contrast, find no correlation between the number of waves and interannual variability.

Gu et al. [62] concluded that there are two distinct relationships between waves and convection. Southerly wind perturbations lag precipitation south of 15° N, but lead it north of 15° N. The relationship can be in-phase or out-of-phase. The relationship shifts as the waves traverse the African continent and also depends on latitude with respect to the AEJ [200]. At 10° N convection occurs in northerly flow east of the Greenwich meridian, then shifts into the wave trough. North of the AEJ, at 15° N, convection remains in the southerly flow throughout it transit across Africa. Some more complex forms of the wave also exist, in which convection has an inverted-V appearance and is wrapped around the trough, which is an area of low wind and clear skies [201]. The lower frequency 6 to 9 day waves tend to have a rainfall maximum in the cyclonic flow to the north and a rainfall minima in the anticyclonic flow to the south [202].

Hall et al. [119] provide a comprehensive review of idealized studies of AEW development in relationship to the African Easterly Jet. They had assumed a zonally uniform basic state. Using a linear primitive equation model, they considered the impact of zonal asymmetries of the AEJ, and showed that these modified the developing dynamic modes and resultant waves. Leroux and Hall [193] similarly showed that the response to intraseasonal variability of the AEJ depends strongly on its basic state. In this case, a convective trigger was utilized. Some configurations of the AEJ failed to produce waves and others produced very strong waves. Requirements for the latter include a strong jet, strong vertical shear, a strong and extended potential vorticity reversal. The strong vertical shear results when the AEJ core is aligned with the maximum of surface westerlies. However, the vertical shear in the upper troposphere appears to be equally important [136].

Grist [198], Grist et al. [144], and Nicholson et al. [136], using both observations and linear dynamic models, also examined the impact of basic state on wave development. In this case, the basic states corresponded to two composites of wet years and two composites of dry years. The four months of June through August were considered. In agreement with Leroux and Hall [193], they demonstrated the need for strong vertical shear to produce strong waves and the presence of this shear when the AEJ core overlies strong surface westerlies. However, it appeared that the major factor responsible for the stronger waves in the wetter years was contrasts in barotropic instability [144]. With the equatorial westerlies extending to the mid-troposhere in wet years, the horizontal shear was particularly strong.

Several other contrasts between the wet and dry cases were noted. These include faster growth rates and phase speeds, stronger waves, and a greater contribution from the longer period waves. In the wet years the waves have a barotropic structure and extend throughout the troposphere. Those of dry years extend only to mid-levels. Two distinct wavelengths are also apparent in wet years, 3 to 4 km and 6 to 7 km. The latter are planetary scale waves, suggesting interactions between the upper troposphere and the surface. This is consistent with the suggestion of Cornforth et al. [112] that interaction with convection weakens the surface manifestation of the waves during the peak monsoon season.

LARGE-SCALE FACTORS IN VARIABILITY: THE ROLE OF SEA-SURFACE TEMPERATURES

Some of the earliest work on the relationship between Sahel rainfall and sea-surface temperatures was prompted by the drought conditions that commenced in the late 1960s. Seminal papers were published by Lamb [203, 204], who suggested links to SSTs in the equatorial and subtropical Atlantic, and by Folland et al. [205], who highlighted the role of inter-hemispheric SST gradients.

Since that time there has been vigorous debate but no definitive answers on the major drivers. Some of the less controversial conclusions are that SST patterns play a critical role in the variability (e.g., [99, 206–208]), that a variety of regional and global SST patterns play a role (e.g., [209, 210]), that the link to SSTs is probably different for high- and low-frequency (i.e., interannual and interdecadal) variability [211–213], that the relationships are not stationary over time (e.g., [104, 213, 214]), and that external forcing via Saharan dust and/or greenhouse gases is probably playing a role in Sahel rainfall variability (e.g., [99, 208,215–217]).

Some of the more controversial issues are the impact of ENSO on the Sahel, the relative importance of the various oceans and the Mediterranean, and the extent to which climate models can adequately simulate Sahel rainfall. There is disagreement concerning the models' ability to simulate mean climatology and the long drying trend since the 1970s, but universal agreement that the models (Figure 21) cannot produce reliable assessments of future climate change in the region

[99, 208, 212, 215, 218–220]. Modeling efforts are hindered by the fact that there are multiple competing physical mechanisms, as described below [219]. The search for a better understanding of Sahel climate variability is complicated by the fact that most studies rely on numerical modeling. Details of the modeling issue are beyond the scope of the review. However, excellent reviews of state-of-art modeling and model intercomparisons can be found in Mohino et al. [214], Patricola and Cook [221], Lau et al. [215], Biasutti et al. [99], Rodriguez-Fonseca et al. [218], Hourdin et al. [222], Druyan et al. [223], Cook et al. [224], Nikulin et al. [225], and Ruti et al. [226].

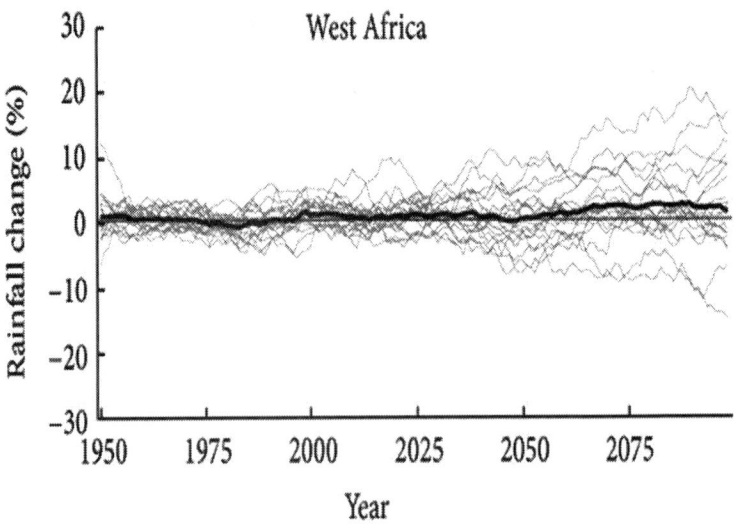

Figure 21: Average of precipitation over West Africa in the IPCC 4AR model simulations: twentieth century simulations from 1950 to 2000, and A1B scenario simulations from 2000 to 2100 (from [99]). Each grey line represents one model, and the thicker black line is the multi-model mean.

For the most part, model simulations of Sahel rainfall have confirmed the importance of the two mechanisms proposed by Lamb [203, 204] and Folland et al. [205]: the role of equatorial Atlantic SSTs and the role of interhemispheric SST gradients. Hoerling et al. [227] emphasized the latter, suggesting that the drying trend over the last half of the twentieth century was due to a warming of the South Atlantic relative to the North Atlantic. Joly et al. [212] implicated the former as a major factor in the high frequency variability of Sahel rain, suggesting that its role is the modulation of the northward migration of the West African monsoon. Losada et al. [228] suggested the mechanism is related to the degree of low-level convergence over the Sahel. In agreement with Hoerling et al., they also found that the interhemispheric mode is linked to the low-frequency drying trend. Sun et al. [229] take the link to the Atlantic further south, demonstrating a link to the Antarctic Oscillation via Atlantic SSTs. This is consistent with the work of Nicholson and Webster [95], who show a strong link to the extratropical South Atlantic.

Others have emphasized the influence of the North Atlantic. Polo et al. [230], for example, demonstrate an influence of the North Atlantic Oscillation. Several studies have suggested a relationship to the Atlantic Multidecadal Oscillation (AMO, also termed Atlantic Meridional Overturning). This is the average temperature over the North Atlantic. The relationship is particularly strong for low-frequency variations [231–233], with strong links extending back to at least the beginning of the nineteenth century (Figure 22) (Dezfuli and Nicholson, in preparation). The paleoclimate work of Shanahan et al. [234] suggests that the relationship holds on even longer time scales.

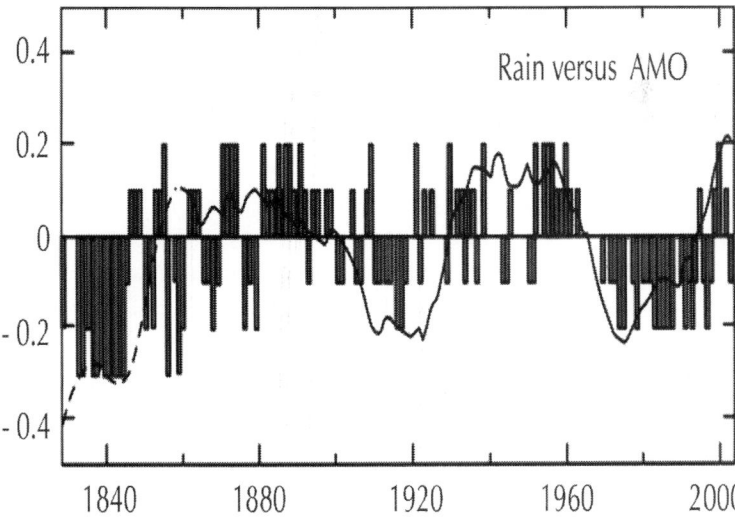

Figure 22: Semiquantitative index of Sahel rainfall since 1823 compared with the AMO. The scale of the latter AMO data has been adjusted to match the scale of the SST-based index (from [100]).

On the other hand, model simulations of Hodson et al. [235] suggest that the AMO is not the primary cause of the recent drying trend in the Sahel. This is consistent with the conclusion of Losada et al. [104] that there has been a change in rainfall patterns over the Sahel in recent decades. Mostly notably, the well-known dipole rainfall pattern, with an opposition between the Sahel and Guinea Coast, seems to have disappeared. The suggested reason for its disappearance is that the influence of SSTs in the Gulf of Guinea, which are linked to the dipole [236], and other areas of the Atlantic has been counteracted by Pacific warming [104,213].

Two of the first papers to emphasize the role of the Pacific and ENSO in modulating Sahel rainfall were Rowell [237] and Janicot and Sultan [65]. Based at least partially on model results, these conclusions were at odds with the conclusions of several observational analyses (e.g., [76, 238–240]). The issue was reconciled by Ward [211], who separately evaluated low- and high-frequency variability and found that ENSO influenced the latter. Further support for a link to ENSO comes from Joly et al. [212] and Losada et al. [104]. Joly and Voldoire noted that the timing of the ENSO events are critical to the teleconnection with Sahel rainfall, which is strongest in the developing phase of el Niño or the decaying phase of some La Niña episodes. Bader and Latif [241] suggest that the Pacific's influence is mainly in the eastern Sahel.

Both model simulations and observations suggest that the Indian Ocean also has a strong influence on Sahel rainfall. This suggests an influence between Sahel rainfall and the Indian monsoon (e.g., [242]). Chung and Ramanathan [243] suggest that both Indian Ocean warming and SST gradients in the Indian Ocean play a role. Bader and Latif [241] conclude that the Indian Ocean was the probable driver of the drying trend in the Sahel from the 1950s to the 1990s. Bader and Latif [244] evaluated the extreme drought year 1983 and concluded that the main factor was Indian Ocean SSTs, showing that they can influence not only decadal scale variability but also interannual. In contrast to most observational studies they suggested that the Atlantic dipole was not the main cause. They also concluded the very intense el Niño of that year had a weak influence at best. Mohino et al. [214] emphasize both Indian and Pacific Ocean anomalies, suggesting they can also, like Gulf of Guinea SSTs, displace the rainbelt latitudinally over the Sahel. Notably, their suggested mechanism is most prominent in August, the month that contributes the most to interannual variability. Several authors have suggested that the probably mechanism of influence via the Indian Ocean is equatorial wave dynamics, especially westward propagating Rossby waves [206, 241, 245, 246]. This conclusion is particularly interesting, because studies of intraseasonal variability of Sahel rainfall and disturbances systems show a strong link to equatorial waves, especially Rossby waves (see Section 6).

One of the first papers to examine the link between the Mediterranean and Sahel rainfall is that of Rowell [209]. His observational analysis suggested that during the period 1947 to 1996 the impact of the Mediterranean was on par with that of the Pacific and only somewhat weaker than that of the Atlantic. Model simulations pointed towards moisture flux as the mechanism of the association. Several modeling studies have similarly noted a link between Mediterranean warm events and anomalously high Sahel rainfall. The observational study of Polo et al. [247] suggests that the most direct influence on Sahel rainfall comes from the Mediterranean, but that this may be a fingerprint of larger scale forcing. The suggested mechanisms of the relationship include enhanced evaporation over the Mediterranean [248], enhancement or northward displacement of the southwesterly monsoon flow [88], enhancement of moisture flux and the Saharan Heat Low [249], and increased convergence [250]. Fontaine et al. [88] conclude that the impact is different for the eastern and western Mediterranean Basin, with the eastern Mediterranean having the greater influence on the Sahel. Observations suggest that warming and cooling have a symmetric impact but a model simulation suggests that only warming has an impact.

The conclusion that one can draw from these various studies is that multiple, competing physical mechanisms regulate interannual and decadal variability in the Sahel [219]. The dominant mechanism varies with location within the Sahel and with the time period in question. For that reason it is difficult to identify the contribution of any particular ocean. Moreover, each model has its biases and these influence the degree to which any particular mechanism can be simulated. Hence, the large disparity of model results.

LAND SURFACE EFFECTS, INCLUDING AEROSOLS

The possibility of the land surface influencing Sahel climate was first addressed by Charney [251]. In response to the occurrence of severe drought in the 1970s, he proposed that overgrazing in the region may have been the cause. The proposed mechanism was a change in surface albedo as a result of the denuding of highly reflective soils. Subsequently other mechanisms were proposed, including such factors as changes in soil moisture and evapotranspiration. These were assumed to accompany a process called "desertification", in which human transformation of the land led to an irreversible decline in productivity.

This stimulated a plethora of model simulations of the impact of such land surface changes (see reviews by Nicholson [102], Entekhabi [252], Giannini et al. [208]). These have established the potential for changes in the land surface over the Sahel to influence the region's climate [253, 254]. In most cases, feedback mechanisms were demonstrated by which an existing drought could be intensified rather than triggered (e.g., [255]). However, observational evidence was lacking. At the same time, numerous studies showed that the extent of desertification is the Sahel had been dramatically exaggerated and was generally a reversible process [256–259].

During the last decade or so the nature of research on land surface feedback on Sahel climate has shifted markedly. The changes include use of regional climate models instead of global models (e.g., [260–262]), emphasis on the impacts on individual convective or synoptic systems as opposed to seasonal or large-scale climate (e.g., [263]), consideration of aerosols and the associated Saharan Air Layer (e.g., [159, 164]), and examination of the hypothesis that vegetation feedback can produce abrupt climate change in the Sahel [264]. Recent research in each of these areas is reviewed.

Land Surface Processes in Regional Models

Perhaps the most important result of the use of regional models to study land surface processes is the uniform consensus that interactive land surface schemes improve the performance of climate models [216,265–267]. They improve the simulation of hydrological processes, circulation features, and enhance predictive capacity. The net result is a confirmation that land-atmosphere interaction is an inherent feature of the mean climatology of the region.

Some of the most promising working on land surface processes is the studies linking them to convection. In semiarid regions the surface fluxes of heat and moisture are strongly impacted by the occurrence of convective rainfall [268], but the so-produced surface anomalies also influence the development of convective rainfall. The relevant characteristic is not the overall state of the land surface, but the spatial heterogeneity of the characteristics of moisture and temperature that play a role (e.g., [269–271]. This conclusion, also reached long ago in early modeling experiments (e.g., [272]), has stood the test of time.

Relationship between Land-Surface Characteristics and Convection

Most of the work on the links between the land surface and convection has been done with model simulations. These have concluded that surface flux anomalies related to soil moisture and temperature can induce secondary circulations that impact boundary layer depth and cloud development and trigger convection [273, 274]. The surface fluxes have an impact on the diurnal cycle of convection [275]. The coupling of the atmosphere and land surface also appears to produce unstable modes that do not exist in the atmosphere-only part of the system [276]. The result is wave disturbances, the propagation of which is governed by the spatial configuration of atmospheric and soil moisture anomalies. In the Sahel rainfall-induced anomalies can persist for several days after a storm, long enough to modify the overlying boundary layer [277]. Boundary layer changes are linked mainly to long length scales of surface heterogeneities [268] because advective effects may override the influence of the land-surface at smaller scales [278]. Squall-line rainfall, on the other hand, shows the strongest response to anomalies on small length scales comparable to that of convection.

Observational studies have produced convincing evidence of the impact of land-surface heterogeneities on convection in the semi-arid

Sahel, particularly the drier northern sector. The first of these was that of Taylor and Lebel [279], who used data from the HAPEX-Sahel field experiment of 1992 [280]. They found that the probability of rainfall at a given location was strongly correlated with that location having received antecedent rainfall from a previous event. In other words, once a location has received significant rainfall, it is more likely than other locations to receive rainfall from subsequent disturbances that pass by.

More recent observational studies have provided more detail into the mechanisms involved in the feedback. Dry anomalies (areas of higher temperature and lower moisture than the surroundings) can be several hundred kilometers and larger. These "hot spots" generate anomalous heat lows during the day and cyclonic vorticity at night [269]. These vortices may influence the characteristics of African easterly waves in the northern Sahel and thereby influence rainfall. These vortices are thus both a cause and effect of the rainfall [281]. Soil moisture anomalies on the scale of 10 to 40 kilometers appear to exert strong control on storm initiation [270]. An interesting case study of such an effect was made for a convective system that occurred over northern Mali during the AMMA Special Observing Period [263]. The impacts of soil moisture patches are evident even at scales of 200 km [282]. The initiation of convection by such patches is favored when the soil inhomogeneity is superimposed upon a convergence zone [283]. However, the process is also affected by other factors, such as humidity in the boundary layer and by convective available potential energy (CAPE).

The impact of soil moisture heterogeneities is strongly dependent on the stage of convection. Convection is often initiated over strong moisture gradients or over the dry "hot spots". However, dry soil appears to weaken mature systems [271] while moisture soil will enhance mature convective systems [263].

One of the still open questions is whether the feedback between the land surface and convection merely modifies the spatial distribution of rainfall or affects the overall amount as well. In modeling studies by Lauwaet et al. [284, 285] the surface changes introduced did not influence the amount of rainfall, despite the changes in surface fluxes and CAPE. Model simulations of Gantner and Kalthoff [271] suggest that the lack of a net impact may be a result of the contrasting impact on developing versus mature convective systems. The triggering of convection was favored by drier surfaces and/or soil moisture inhomogeneities, but mature systems weakened when they approached drier surfaces. Gaertner et al. [286] found that introducing a soil moisture perturbation into a model had different effects at different scales: a reduction in rainfall over the wet path itself but a net increase in rainfall over a larger area. Moufouma-Okia and Rowell [287], using a regional climate model, found that soil moisture anomalies had little impact on the overall amount of rainfall, creating only small and random intraseasonal, interannual and spatial variations. However, they noted that the conclusion may be model dependent or influenced by the initial soil moisture data introduced into the model. Hence the question is very open.

Impact of Aerosols

The origin of the dust layer over West Africa is fine material eroded from the dry remnants of Holocene lakes that once spread across the Sahel [160]. African Easterly Waves play an important role in the mobilization of the dust, although extratropical disturbances also play a role in the western Sahel [288]. Some 20% of the entrainment into the atmosphere over North Africa is linked to AEWs [289]. The dust-loading over the Sahel increased tremendously between the 1950s and the 1980s, a consequence of the long period of drought [290].

The main impact of the dust is on the atmospheric radiation budget. However, the particulates also serve as ice nuclei and cloud condensation nuclei [291–293]. The radiative effects depend on the composition of the dust, that is, on its relative role in absorbing versus reflecting radiation (see review in [102]). Hence the radiative effects are different for the mineral-based Saharan dust and the smoke introduced by biomass burning [294]. For the most part, the dust acts as an "elevated heat pump" [215], warming the mid-troposphere but cooling the surface [295, 296]. Maximum heating is in the layer 850 mb to 700 mb [294].

The dust appears to intensify the African Easterly Jet, consistent with the mid-level heating, and weaken the Tropical Easterly Jet [216]. Dust outbreaks also affect the Saharan Heat Low. The radiative effect increases its thickness, but the impact of the dust on the AEJ and AEWs affects the heat low. Most studies conclude that the net effect is to reduce precipitation over the Sahel [296, 297], particularly that associated with deep convection [294]. Lau et al. [215], however, suggest that the dust can strength the monsoon and shift it northward. This implies an increase in rainfall over the Sahel.

Abrupt Climate Change

An outgrowth of the research on land-surface processes has been the hypothesis that these processes can lead to abrupt climate change. The issue has been related primarily to the question of the creation of the Sahara Desert, but it has implications for the Sahel as well [298]. Reviews are provided by Rietkerk et al. [264] and Kabat et al. [299].

This hypothesis results from contemporaneous findings in climatology, paleoclimatology, and ecology. A dust core in the Atlantic just off the West African coast suggested that, in the Sahara, the transition from savanna to desert which took place a few thousand years ago may have occurred within a few hundred years [300]. While that finding contradicts the more conventional view of the termination of the African humid period (e.g., [301]), climate models that included interactive vegetation were able to replicate such an abrupt transition [302, 303]. These models have also shown an accompanying feedback of the vegetation on precipitation.

At the same time, ecological theory was suggesting that vegetation-climate-nutrient feedbacks were such that in arid and semi-arid regions the vegetation is self-organizing [304, 305]. A corollary to this is that very rapid transitions from wet and vegetated ground to dry and bare ground could occur in the presence of an external climate trigger [306, 307]. The relevant feedbacks with vegetation cover include soil moisture, shading, and nutrient retention [308–312]. When critical thresholds in the climate-vegetation system are surpassed, catastrophic shifts in the vegetation state can occur.

SUMMARY

Rainfall in the Sahel has shown some degree of recovery since the extreme dry episode of the 1970s and 1980s. However, certain characteristics of the rainfall regime appear to have changed. There is less spatial coherence and less temporal persistence. The contrast between conditions in the eastern and western Sahel is becoming increasingly stronger. The peak month appears to have shifted from August to July.

Most recent work on the Sahel has been in the context of the West African monsoon overall. New paradigms for the monsoon and associated ITCZ have emerged. These emphasize features in the upper atmosphere, as well as the Saharan Heat Low. Feedback mechanisms have also been emphasized, especially the coupling of convection with atmospheric dynamics. Many researchers are arguing the viewpoint that convection is a trigger/driver, as opposed to a passive response to atmospheric dynamics. The role of land surface effects on convection is also becoming increasingly demonstrated. The role of tropical/extratropical interaction has also been brought to the forefront during the last decade by way of studies of hybrid systems such as diagonal troughs and evidence of the role of the Mediterranean in modulating Sahel rainfall.

An interesting result that emerges from recent research is the contrast in the Sahel rainfall regime during the premonsoon season and the peak monsoon season. During the former (May to late June/early July), the tropical rainbelt lies south of the Sahel, around 5° N. During the peak monsoon season the rainbelt is centered around 10° N. Lebel et al. [22] term these the ocean and continental regimes, respectively. The shift between the two locations/regimes is abrupt (the "monsoon jump").

It has long been established that interannual and decadal variability in the Sahel is related primarily to anomalies in the months of August and September. The long drying trend is clearly linked to conditions in those months. This is considered to be the peak monsoon season, as well as the period of dominance of the "continental" regime. Recent studies have emphasized how the precipitation character and its dynamic controls differ during the continental regime and the preceding premonsoon season/ocean regime. With the shift between the ocean and continental regimes, the horizontal wind shear over the Sahel increases markedly, the vertical shear decreases, and the AEJ weakens. The contribution of stratiform rain increases, rainfall becomes less intense, and fewer squall lines occur. The influence of the Indian and Pacific Oceans appears to be strongest during the continental regime than during the ocean regime. At this time, wave activity is more closely coupled to convection and to squalls. The relationship of rainfall to the surface ITCZ becomes weaker (it is weakest in August). Longer period waves (6.25 to 7.5 days) also provide a greater contribution at this time.

Many studies have examined the relationship between the Sahel and large-scale processes. Very new results include the roles played by the Madden-Julian Oscillation and by equatorial waves. These are tied into global sea-surface temperature fluctuations. In the past, emphasis has been placed on the Atlantic Ocean in the context of interannual variability over the Sahel. Both observations and numerical simulations identify contributions of each of the global oceans.

The oceans manifest their influence in several ways, but most markedly in changes in several circulation features: the Tropical Easterly Jet, the African Easterly Jet, the low-level African Westerly Jet, and the Saharan Heat Low. The result is two major modes of rainfall variability, one associated with a latitudinal displacement of the tropical rainbelt over West Africa and one associated with a general intensification or weakening of the rainbelt. These modes are associated with different dynamic states and their existence and contrasts likely play a role in the sometimes contradictory conclusions that have been drawn concerning the role of the oceans. Wet or dry conditions in the Sahel can be associated with either mode. However the common factors in producing wet conditions appear to be pressure and temperature patterns over the Atlantic and the intensity of the Tropical Easterly Jet over West Africa.

REFERENCES

1. S. E. Nicholson and J. P. Grist, "The seasonal evolution of the atmospheric circulation over West Africa and equatorial Africa," Journal of Climate, vol. 16, pp. 1013–1030, 2003.

2. A. Ali and T. Lebel, "The Sahelian standardized rainfall index revisited," International Journal of Climatology, vol. 29, no. 12, pp. 1705–1714, 2009.

3. S. Nicholson, "On the question of the "recovery" of the rains in the West African Sahel," Journal of Arid Environments, vol. 63, no. 3, pp. 615–641, 2005. · ·

4.   P. J. Lamb, "Sub-Saharan rainfall update for 1982: continued drought," Journal of Climatology, vol. 3, no. 4, pp. 419–422, 1983.

5.   P. J. Lamb and R. A. Peppler, "Further case studies of tropical Atlantic surface atmospheric and oceanic patterns associated with sub-Saharan drought," Journal of Climate, vol. 5, pp. 476–488, 1992.

6.   S. E. Nicholson and I. M. Palao, "A re-evaluation of rainfall variability in the Sahel. Part I. Characteristics of rainfall fluctuations," International Journal of Climatology, vol. 13, no. 4, pp. 371–389, 1993.

7.   S. E. Nicholson, "The nature of rainfall fluctuations in subtropical West Africa (Guinea Sahel Soudan)," Monthly Weather Review, vol. 108, no. 4, pp. 473–487, 1980.

8.   T. Lebel and A. Ali, "Recent trends in the Central and Western Sahel rainfall regime (1990–2007),"Journal of Hydrology, vol. 375, no. 1-2, pp. 52–64, 2009. · ·

9.   S. Janicot, C. D. Thorncroft, A. Ali et al., "Large-scale overview of the summer monsoon over West Africa during the AMMA field experiment in 2006," Annales Geophysicae, vol. 26, no. 9, pp. 2569–2595, 2008.

10.   J. L. Redelsperger, C. D. Thorncroft, A. Diedhiou, T. Lebel, D. J. Parker, and J. Polcher, "African Monsoon Multidisciplinary Analysis: an international research project and field campaign," Bulletin of the American Meteorological Society, vol. 87, no. 12, pp. 1739–1746, 2006. · ·

11.   A. Boone, P. De Rosnay, G. Balsamo et al., "The AMMA land surface model intercomparison project (ALMIP)," Bulletin of the American Meteorological Society, vol. 90, no. 12, pp. 1865–1880, 2009. · ·

12.   T. Lebel, B. Cappelaere, S. Galle et al., "AMMA-CATCH studies in the Sahelian region of West-Africa: an overview," Journal of Hydrology, vol. 375, no. 1-2, pp. 3–13, 2009. · ·

13.   C. D. Thorncroft, D. J. Parker, and R. R. Burton, "The JET2000 project—aircraft observations of the African easterly jet and African easterly waves," Bulletin of the American Meteorological Society, vol. 84, pp. 337–351, 2003.

14. J. F. Griffiths, World Survey of Climatology, vol. 10, Elsevier, 1972.

15. C. Zhang, P. Woodworth, and G. Gu, "The seasonal cycle in the lower troposphere over West Africa from sounding observations," Quarterly Journal of the Royal Meteorological Society, vol. 132, no. 621, pp. 2559–2582, 2006. · ·

16. B. J. Mason, "The GARP Atlantic tropical experiment," Nature, vol. 255, no. 5503, pp. 17–20, 1975. · ·

17. Y. Y. Yan, "Intertropical Convergence Zone (ITCZ)," in Encyclopedia of World Climatology, J. E. Oliver, Ed., pp. 429–432, 2005.

18. R. L. Miller, "The intertropical convergence zone," in Encyclopedia of Climate and Weather, S. H. Schneider, Ed., vol. 1, pp. 445–448, 1996.

19. J. R. Holton, J. M. Wallace, and J. A. Young, "On boundary layer dynamics and the ITCZ," Journal of the Atmospheric Sciences, vol. 28, pp. 275–280, 1971.

20. S. E. Nicholson, "A revised picture of the structure of the "monsoon" and land ITCZ over West Africa," Climate Dynamics, vol. 32, no. 7-8, pp. 1155–1171, 2009. · ·

21. R. A. Tomas, J. R. Holton, and P. J. Webster, "The influence of cross-equatorial pressure gradients on the location of near-equatorial convection," Quarterly Journal of the Royal Meteorological Society, vol. 125, no. 556, pp. 1107–1127, 1999.

22. T. Lebel, A. Diedhiou, and H. Laurent, "Seasonal cycle and interannual variability of the Sahelian rainfall at hydrological scales," Journal of Geophysical Research D, vol. 108, no. 8, pp. 14–11, 2003.

23. S. W. Nesbitt and E. J. Zipser, "The diurnal cycle of rainfall and convective intensity according to three years of TRMM measurements," Journal of Climate, vol. 16, no. 10, pp. 1456–1475, 2003.

24. P. Peyrillé, J. P. Lafore, and J. L. Redelsperger, "An idealized two-dimensional framework to study the West African Monsoon—part I: validation and key controlling factors," Journal of the Atmospheric Sciences, vol. 64, no. 8, pp. 2765–2782, 2007. · ·

25. C. D. Thorncroft, H. Nguyen, C. Zhang, and P. Peyrille, "Annual cycle of the West African monsoon: regional circulations and

associated water vapour transport," Quarterly Journal of the Royal Meteorological Society, vol. 137, no. 654, pp. 129–147, 2011. · ·

26. C. Lavaysse, C. Flamant, and S. Janicot, "Regional-scale convection patterns during strong and weak phases of the Saharan heat low," Atmospheric Science Letters, vol. 11, no. 4, pp. 255–264, 2010. · ·

27. P. Peyrillé and J. P. Lafore, "An idealized two-dimensional framework to study the West African Monsoon—part II: large-scale advection and the diurnal cycle," Journal of the Atmospheric Sciences, vol. 64, no. 8, pp. 2783–2803, 2007. · ·

28. G. J. Gu and R. F. Adler, "Seasonal evolution and variability associated with the West African monsoon system," Journal of Climate, vol. 17, pp. 3364–3377, 2004.

29. D. J. Parker, R. R. Burton, A. Diongue-Niang, et al., "The diurnal cycle of the West African monsoon circulation," Quarterly Journal of the Royal Meteorological Society, vol. 131, pp. 2839–2860, 2005.

30. J. P. Grist and E. Nicholson, "A study of the dynamic factors influencing the rainfall variability in the West African Sahel," Journal of Climate, vol. 14, no. 7, pp. 1337–1359, 2001.

31. C. D. Thorncroft and M. Blackburn, "Maintenance of the African easterly jet," Quarterly Journal of the Royal Meteorological Society, vol. 125, no. 555, pp. 763–786, 1999.

32. D. S. Nolan, C. Zhang, and S. H. Chen, "Dynamics of the shallow meridional circulation around intertropical convergence zones," Journal of the Atmospheric Sciences, vol. 64, no. 7, pp. 2262–2285, 2007. · ·

33. C. Zhang, D. S. Nolan, C. D. Thorncroft, and H. Nguyen, "Shallow meridional circulations in the tropical atmosphere," Journal of Climate, vol. 21, no. 14, pp. 3453–3470, 2008. · ·

34. F. Couvreux, F. Guichard, O. Bock, B. Campistron, J. P. Lafore, and J. L. Redelsperger, "Synoptic variability of the monsoon flux over West Africa prior to the onset," Quarterly Journal of the Royal Meteorological Society, vol. 136, no. 1, pp. 159–173, 2010. · ·

35. J. Cuesta, C. Lavaysse, C. Flamant, M. Mimouni, and P. Knippertz, "Northward bursts of the West African monsoon leading to rainfall

over the Hoggar Massif, Algeria," Quarterly Journal of the Royal Meteorological Society, vol. 136, no. 1, pp. 174–189, 2010. · ·

36. S. E. Nicholson, "The intensity, location and structure of the tropical rainbelt over west Africa as factors in interannual variability," International Journal of Climatology, vol. 28, no. 13, pp. 1775–1785, 2008. · ·

37. S. E. Nicholson, "On the factors modulating the intensity of the tropical rainbelt over West Africa,"International Journal of Climatology, vol. 29, no. 5, pp. 673–689, 2009. · ·

38. P. Laux, H. Kunstmann, and A. Bárdossy, "Predicting the regional onset of the rainy season in West Africa," International Journal of Climatology, vol. 28, no. 3, pp. 329–342, 2008. · ·

39. B. Sultan and S. Janicot, "The West African monsoon dynamics. Part II: The"preonset" and "onset" of the summer monsoon," Journal of Climate, vol. 16, pp. 3407–3427, 2003.

40. R. Marteau, V. Moron, and N. Philippon, "Spatial coherence of Monsoon onset over Western and Central Sahel (1950–2000)," Journal of Climate, vol. 22, no. 5, pp. 1313–1324, 2009. · ·

41. B. Sultan and S. Janicot, "Abrupt shift of the ITCZ over West Africa and intra-seasonal variability,"Geophysical Research Letters, vol. 27, no. 20, pp. 3353–3356, 2000. · ·

42. P. Drobinski, S. Bastin, S. Janicot et al., "On the late northward propagation of the west African monsoon in summer 2006 in the region of Niger/Mali," Journal of Geophysical Research D, vol. 114, no. 9, Article ID D09108, 2009. · ·

43. O. Dieng, P. Roucou, and S. Lovet, "Intra-seasonal variability of precipitation in Senegal (1951–1996)," Secheresse, vol. 19, pp. 87–93, 2008.

44. G. A. Dalu, M. Gaetani, and M. Baldi, "A hydrological onset and withdrawal index for the West African monsoon," Theoretical and Applied Climatology, vol. 96, no. 1-2, pp. 179–189, 2009. · ·

45. E. Flaounas, S. Janicot, S. Bastin, and R. Roca, "The West Africa monsoon onset in 2006: sensitivity to surface albedo, orography, SST and synoptic scale dry-air intrusions using WRF," Climate Dynamics, vol. 38, pp. 685–708, 2012.

46. E. Flaounas, S. Janicot, S. Bastin, R. Roca, and E. Mohino, "The role of the Indian monsoon onset in the West African monsoon

onset: observations and AGCM nudged simulations," Climate Dynamics, vol. 38, no. 5-6, pp. 965–983, 2011. · ·

47. P. Camberlin, B. Fontaine, S. Louvet, P. Oettli, and P. Valimba, "Climate adjustments over Africa accompanying the Indian monsoon onset," Journal of Climate, vol. 23, no. 8, pp. 2047–2064, 2010. · ·

48. C. Schumacher and R. A. House, "Stratiform precipitation production over sub-Saharan Africa and the tropical East Atlantic as observed by TRMM," Quarterly Journal of the Royal Meteorological Society, vol. 132, no. 620, pp. 2235–2255, 2006. · ·

49. L. Le Barbé, T. Lebel, and D. Tapsoba, "Rainfall variability in West Africa during the years 1950–90,"Journal of Climate, vol. 15, no. 2, pp. 187–202, 2002.

50. B. Sultan, S. Janicot, and A. Diedhiou, "The West African monsoon dynamics. Part I: documentation of interaseasonal variability," Journal of Climate, vol. 21, pp. 3389–3406, 2003.

51. S. Sijikumar, P. Roucou, and B. Fontaine, "Monsoon onset over Sudan-Sahel: simulation by the regional scale model MM5," Geophysical Research Letters, vol. 33, no. 3, Article ID L03814, 2006. · ·

52. R. Ramel, H. Gallé, and C. Messager, "On the northward shift of the West African monsoon," Climate Dynamics, vol. 26, no. 4, pp. 429–440, 2006. · ·

53. Y. Okumura and S. P. Xie, "Interaction of the Atlantic equatorial cold tongue and the African monsoon," Journal of Climate, vol. 17, pp. 3589–3602, 2004.

54. S. M. Hagos and K. H. Cook, "Dynamics of the West African monsoon jump," Journal of Climate, vol. 20, no. 21, pp. 5264–5284, 2007. · ·

55. A. Sealy, G. S. Jenkins, and S. C. Walford, "Seasonal/regional comparisons of rain rates and rain characteristics in West Africa using TRMM observations," Journal of Geophysical Research D, vol. 108, no. 10, pp. 3–21, 2003.

56. M. A. Bell and P. J. Lamb, "Integration of weather system variability to multidecadal regional climate change: the West African Sudan-Sahel zone, 1951–98," Journal of Climate, vol. 19, no. 20, pp. 5343–5365, 2006. · ·

57. S. E. Nicholson, Dryland Climatology, Cambridge University Press, Cambridge, 2011.

58. M. Le Lay and S. Galle, "Seasonal cycle and interannual variability of rainfall at hydrological scales. The West African monsoon in a Sudanese climate," Hydrological Sciences Journal, vol. 50, no. 3, pp. 509–524, 2005. · ·

59. E. Mohino, S. Janicot, H. Douville, and L. Z. X. Li, "Impact of the Indian part of the summer MJO on West Africa using nudged climate simulations," Climate Dynamics, vol. 38, pp. 2319–2334, 2012.

60. S. Janicot, F. Mounier, N. M. J. Hall, S. Leroux, B. Sultan, and G. N. Kiladis, "Dynamics of the West African monsoon. Part IV: analysis of 25–90-day variability of convection and the role of the Indian monsoon," Journal of Climate, vol. 22, no. 6, pp. 1541–1565, 2009. · ·

61. S. Janicot, F. Mounier, S. Gervois, B. Sultan, and G. N. Kiladis, "The dynamics of the West African monsoon—part V: the detection and role of the dominant modes of convectively coupled equatorial Rossby waves," Journal of Climate, vol. 23, no. 14, pp. 4005–4024, 2010. · ·

62. G. Gu, R. F. Adler, G. J. Huffman, and S. Curtis, "African easterly waves and their association with precipitation," Journal of Geophysical Research D, vol. 109, no. 4, Article ID D04101, 12 pages, 2004.

63. S. L. Lavender and A. J. Matthews, "Response of the West African monsoon to the Madden-Julian oscillation," Journal of Climate, vol. 22, no. 15, pp. 4097–4116, 2009. · ·

64. M. J. Ventrice, C. D. Thorncroft, and P. E. Roundy, "The madden-julian oscillation's influence on african easterly waves and downstream tropical cyclogenesis," Monthly Weather Review, vol. 139, pp. 2704–2722, 2011.

65. S. Janicot and B. Sultan, "Intra-seasonal modulation of convection in the West African monsoon," Geophysical Research Letters, vol. 28, no. 3, pp. 523–526, 2001. · ·

66. F. Mounier, S. Janicot, and G. N. Kiladis, "The west African monsoon dynamics. Part III: the quasi-biweekly zonal dipole," Journal of Climate, vol. 21, no. 9, pp. 1911–1928, 2008. · ·

67. E. D. Maloney and J. Shaman, "Intraseasonal variability of the West African monsoon and Atlantic ITCZ," Journal of Climate, vol. 21, no. 12, pp. 2898–2918, 2008. · ·

68. S. Leroux, N. M. J. Hall, and G. N. Kiladis, "A climatological study of transient-mean-flow interactions over West Africa," Quarterly Journal of the Royal Meteorological Society, vol. 136, no. 1, pp. 397–410, 2010. · ·

69. C. D. Thorncroft, N. M. J. Hall, and G. N. Kiladis, "Three-dimensional structure and dynamics of African easterly waves. Part III: genesis," Journal of the Atmospheric Sciences, vol. 65, no. 11, pp. 3596–3607, 2008. · ·

70. R. Roehrig, F. Chauvin, and J. P. Lafore, "10–25 day intraseasonal variability of convection over the Sahel: a role of the Saharan Heat Low and midlatitudes," Journal of Climate, vol. 24, pp. 5863–5878, 2011.

71. F. Chauvin, R. Roehrig, and J. P. Lafore, "Intraseasonal variability of the saharan heat low and its link with midlatitudes," Journal of Climate, vol. 23, no. 10, pp. 2544–2561, 2010. · ·

72. E. K. Vizy and K. H. Cook, "A mechanism for African monsoon breaks: mediterranean cold air surges," Journal of Geophysical Research D, vol. 114, no. 1, Article ID D01104, 2009. · ·

73. R. Roca, J. P. Lafore, C. Piriou, and J. L. Redelsperger, "Extratropical dry-air intrusions into the West African monsoon midtroposphere: an important factor for the convective activity over the Sahel,"Journal of the Atmospheric Sciences, vol. 62, no. 2, pp. 390–407, 2005. · ·

74. S. E. Nicholson, "The nature of rainfall variability in Africa south of the equator," Journal of Climatology, vol. 6, no. 5, pp. 515–530, 1986.

75. S. E. Nicholson, "Revised rainfall series for the West African subtropics," Monthly Weather Review, vol. 107, no. 5, pp. 620–623, 1979.

76. S. E. Nicholson, B. Some, and B. Kone, "An analysis of recent rainfall conditions in West Africa, including the rainy seasons of the 1997 El Nino and the 1998 La Nina years," Journal of Climate, vol. 13, no. 14, pp. 2628–2640, 2000.

77. M. D. Dennett, J. Elston, and J. A. Rodgers, "A reappraisal of rainfall trends in the Sahel," Journal of Climatology, vol. 5, no. 4, pp. 353–361, 1985.

78. S. E. Nicholson, "Climatic and environmental change in Africa during the last two centuries," Climate Research, vol. 17, no. 2, pp. 123–144, 2001.

79. Y. L›Hote, G. Mahe, and B. Some, "The 1990s rainfall in the Sahel: the third driest decade since the beginning of the century," Hydrological Sciences Journal, vol. 48, no. 3, pp. 493–496, 2003. · ·

80. A. Dai, P. J. Lamb, K. E. Trenberth, M. Hulme, P. D. Jones, and P. Xie, "The recent Sahel drought is real," International Journal of Climatology, vol. 24, no. 11, pp. 1323–1331, 2004. · ·

81. Y. L›Hôte, G. Mahé, B. Somé, and J. P. Triboulet, "Analysis of a Sahelian annual rainfall index from 1896 to 2000; the drought continues," Hydrological Sciences Journal, vol. 47, no. 4, pp. 563–572, 2002.

82. S. E. Nicholson, A. K. Dezfuli, and D. Klotter, "A two-century precipitation data set for the continent of Africa," Bulletin of the American Meteorological Society, vol. 93, pp. 1219–1231, 2012.

83. S. E. Nicholson, D. Klotter, and A. K. Dezfuli, "Spatial reconstruction of semi-quantitative precipitation fields over Africa during the nineteenth century from documentary evidence and gauge data," Quaternary Research, vol. 78, pp. 13–23, 2012.

84. H. Faure and J. Y. Gac, "Will the Sahelian drought end in 1985?" Nature, vol. 291, no. 5815, pp. 475–478, 1981. · ·

85. J. M. Prospero and T. N. Carlson, "Vertical and areal distribution of saharan dust over western equatorial north-atlantic ocean," Journal of Geophysical Research, vol. 77, pp. 5255–5265, 1972.

86. P. Ozer, M. Erpicum, G. Demarée, and M. Vandiepenbeeck, "The Sahelian drought may have ended during the 1990s," Hydrological Sciences Journal, vol. 48, no. 3, pp. 489–492, 2003. · ·

87. S. Hastenrath and D. Polzin, "Long-term variations of circulation in the tropical Atlantic sector and Sahel rainfall," International Journal of Climatology, vol. 31, no. 5, pp. 649–655, 2011. · ·

88. B. Fontaine, P. Roucou, M. Gaetani, and R. Marteau, "Recent changes in precipitation, ITCZ convection and northern tropical circulation over North Africa (1979–2007)," International Journal of Climatology, vol. 31, no. 5, pp. 633–648, 2011. · ·

89. G. Mahé and J. E. Paturel, "1896–2006 Sahelian annual rainfall variability and runoff increase of Sahelian Rivers," Comptes Rendus, vol. 341, no. 7, pp. 538–546, 2009. · ·

90. L. Olsson, L. Eklundh, and J. Ardö, "A recent greening of the Sahel—trends, patterns and potential causes," Journal of Arid Environments, vol. 63, no. 3, pp. 556–566, 2005. · ·

91. S. M. Herrmann and C. F. Hutchinson, "The changing contexts of the desertification debate," Journal of Arid Environments, vol. 63, no. 3, pp. 538–555, 2005. · ·

92. S. M. Herrmann, A. Anyamba, and C. J. Tucker, "Recent trends in vegetation dynamics in the African Sahel and their relationship to climate," Global Environmental Change, vol. 15, no. 4, pp. 394–404, 2005. · ·

93. J. W. Seaquist, L. Olsson, J. Ardö, and L. Eklundh, "Broad-scale increase in NPP quantified for the African Sahel, 1982–1999," International Journal of Remote Sensing, vol. 27, no. 22, pp. 5115–5122, 2006. · ·

94. S. E. Nicholson, B. Some, J. McCollum et al., et al., "Validation of TRMM and other rainfall estimates with a high-density gauge dataset for West Africa—part II: validation of TRMM rainfall products,"Journal of Applied Meteorology, vol. 42, pp. 1355–1368, 2003.

95. S. E. Nicholson and P. J. Webster, "A physical basis for the interannual variability of rainfall in the Sahel," Quarterly Journal of the Royal Meteorological Society, vol. 133, no. 629, pp. 2065–2084, 2007. · ·

96. S. W. Nesbitt , et al., 2000.

97. S. E. Nicholson, "Rainfall and atmospheric circulation during drought periods and wetter years in West Africa," Monthly Weather Review, vol. 109, no. 10, pp. 2191–2208, 1981.

98. H. Flohn, "Tropische zirkulationsformen im lichte der satellitenaufnahmen," Bonner Meteorologishe Abhandlungen, vol. 21, 82 pages, 1975.

99. M. Biasutti, I. M. Held, A. H. Sobel, and A. Giannini, "SST forcings and Sahel rainfall variability in simulations of the twentieth and twenty-first centuries," Journal of Climate, vol. 21, no. 14, pp. 3471–3486, 2008. · ·

100. A. K. Dezfuli and S. E. Nicholson, "Re-examination of the relationship between the AMO and Sahel rainfall and the potential for long-term forecasting," In press.

101. N. A. Elagib and M. M. Elhag, "Major climate indicators of ongoing drought in Sudan," Journal of Hydrology, vol. 409, pp. 612–625, 2011.

102. S. Nicholson, "Land surface processes and Sahel climate," Reviews of Geophysics, vol. 38, no. 1, pp. 117–139, 2000. · ·

103. S. E. Nicholson and J. P. Grist, "A conceptual model for understanding rainfall variability in the West African Sahel on interannual and interdecadal timescales," International Journal of Climatology, vol. 21, no. 14, pp. 1733–1757, 2001. · ·

104. T. Losada, B. Rodríguez-Fonseca, E. Mohino, J. Bader, S. Janicot, and C. R. Mechoso, "Tropical SST and Sahel rainfall: a non-stationary relationship," Geophysical Research Letters, vol. 39, Article ID L12705, 2012.

105. M. I. Lélé and P. J. Lamb, "Variability of the Intertropical Front (ITF) and rainfall over the West African Sudan-Sahel zone," Journal of Climate, vol. 23, no. 14, pp. 3984–4004, 2010. · ·

106. M. Balme, T. Lebel, and A. Amani, "Dry years and wet years in the Sahel: quo vadimus?" Hydrological Sciences Journal, vol. 51, no. 2, pp. 254–271, 2006. · ·

107. F. Frappart, P. Hiernaux, F. Guichard et al., "Rainfall regime across the Sahel band in the Gourma region, Mali," Journal of Hydrology, vol. 375, no. 1-2, pp. 128–142, 2009. · ·

108. R. A. Bryson, "Drought in Sahelia. Who or what is to blame?" Ecologist, vol. 3, no. 10, pp. 366–371, 1973.

109. E. B. Kraus, "Subtropical droughts and cross-equatorial energy transports," Monthly Weather Review, vol. 105, pp. 1009–1018, 1977.

110. M. K. Miles and C. K. Follard, "Changes in the latitude of the climatic zones of the Northern Hemisphere," Nature, vol. 252, no. 5484, article 616, 1974. · ·

111. T. C. Chen, "Maintenance of the midtropospheric North African summer circulation; Saharan high and African easterly jet," Journal of Climate, vol. 18, no. 15, pp. 2943–2962, 2005. ··

112. R. J. Cornforth, B. J. Hoskins, and C. D. Thorncroft, "The impact of moist processes on the African Easterly Jet-African Easterly Wave system," Quarterly Journal of the Royal Meteorological Society, vol. 135, no. 641, pp. 894–913, 2009. ··

113. A. M. Tompkins, C. Cardinali, J. J. Morcrette, and M. Rodwell, "Influence of aerosol climatology on forecasts of the African Easterly Jet," Geophysical Research Letters, vol. 32, no. 10, Article ID L10801, 2005. ··

114. K. H. Cook, "Generation of the African easterly jet and its role in determining West African precipitation," Journal of Climate, vol. 12, no. 5, pp. 1165–1184, 1999.

115. M. C. R. Kalapureddy, M. Lothon, B. Campistron, F. Lohou, and F. Saïd, "Wind profiler analysis of the African Easterly Jet in relation with the boundary layer and the Saharan heat-low," Quarterly Journal of the Royal Meteorological Society, vol. 136, no. 1, pp. 77–91, 2010. ··

116. R. W. Burpee, "Origin and structure of easterly waves in lower troposphere of North Africa," Journal of the Atmospheric Sciences, vol. 29, pp. 77–90, 1972.

117. J. S. Hsieh and K. H. Cook, "Generation of African easterly wave disturbances: relationship to the African easterly jet," Monthly Weather Review, vol. 133, no. 5, pp. 1311–1327, 2005. ··

118. J. S. Hsieh and K. H. Cook, "On the instability of the African easterly jet and the generation of African waves: reversals of the potential vorticity gradient," Journal of the Atmospheric Sciences, vol. 65, no. 7, pp. 2130–2151, 2008. ··

119. N. M. J. Hall, G. N. Kiladis, and C. D. Thorncroft, "Three-dimensional structure and dynamics of African easterly waves. Part II: dynamical modes," Journal of the Atmospheric Sciences, vol. 63, no. 9, pp. 2231–2245, 2006. ··

120. A. K. Dezfuli and S. E. Nicholson, "A note on long-term variations of the African easterly jet," International Journal of Climatology, vol. 31, pp. 2049–2054, 2011.

121. Z. Wang and R. L. Elsberry, "Modulation of the African easterly jet by a mesoscale convective system," Atmospheric Science Letters, vol. 11, pp. 169–174, 2010.

122. M. V. Ratnam, M. R. Raman, S. K. Mehta et al., "Sub-daily variations observed in Tropical Easterly Jet (TEJ) streams," Journal of Atmospheric and Solar-Terrestrial Physics, vol. 73, no. 7-8, pp. 731–740, 2011. · ·

123. M. R. Raman, V. V. M. Jagannadha Rao, M. Venkat Ratnam et al., "Characteristics of the Tropical Easterly Jet: long-term trends and their features during active and break monsoon phases," Journal of Geophysical Research D, vol. 114, no. 19, Article ID D19105, 2009. · ·

124. D. R. Pattanaik and V. Satyan, "Fluctuations of Tropical Easterly Jet during contrasting monsoons over India: a GCM study," Meteorology and Atmospheric Physics, vol. 75, no. 1-2, pp. 51–60, 2000.

125. V. Sathiyamoorthy, "Large scale reduction in the size of the Tropical Easterly Jet," Geophysical Research Letters, vol. 32, no. 14, Article ID L14802, 2005. · ·

126. M. Hulme and N. Tosdevin, "The Tropical easterly Jet and Sudan rainfall: a review," Theoretical and Applied Climatology, vol. 39, no. 4, pp. 179–187, 1989. · ·

127. Z. T. Segele, P. J. Lamb, and L. M. Leslie, "Large-scale atmospheric circulation and global sea surface temperature associations with horn of Africa June–September rainfall," International Journal of Climatology, vol. 29, no. 8, pp. 1075–1100, 2009. · ·

128. B. R. Srinivasa Rao, D. V. Bhaskar Rao, and V. Brahmananda Rao, "Decreasing trend in the strength of Tropical Easterly Jet during the Asian summer monsoon season and the number of tropical cyclonic systems over Bay of Bengal," Geophysical Research Letters, vol. 31, no. 14, pp. L141031–L141033, 2004. ·

129. H. Flohn, "The tropical easterly jet," Bonner meteorologische Abhandlungen, vol. 4, pp. 1–69, 1964.

130. H. Besler, "The Tropical Easterly Jet as a cause for intensified aridity in the Sahara," Palaeoecology of Africa, vol. 16, pp. 163–172, 1984.

131. P. J. Webster and J. Fasullo, "Monsoon: dynamical theory," in Encyclopedia of Atmospheric Sciences, J. Holton and J. A. Curry, Eds., pp. 1370–1385, Academic Press, London, UK, 2003.

132. T. C. Chen and H. van Loon, "Interannual variation of the tropical easterly jet," Monthly Weather Review, vol. 115, no. 8, pp. 1739–1759, 1987.

133. S. K. Mishra, "Nonlinear barotropic instability of upper-tropospheric tropical easterly jet on the sphere," Journal of the Atmospheric Sciences, vol. 50, no. 21, pp. 3541–3552, 1993.

134. S. K. Mishra and M. K. Tandon, "A combined barotropic-baroclinic instability study of the upper tropospheric tropical easterly jet," Journal of the Atmospheric Sciences, vol. 40, no. 11, pp. 2708–2723, 1983.

135. S. E. Nicholson, A. I. Barcilon, M. Challa, and J. Baum, "Wave activity on the tropical easterly jet,"Journal of the Atmospheric Sciences, vol. 64, no. 7, pp. 2756–2763, 2007. · ·

136. S. E. Nicholson, A. I. Barcilon, and M. Challa, "An analaysis of west African dynamics using a linearized GCM," Journal of the Atmospheric Sciences, vol. 65, no. 4, pp. 1182–1203, 2008. · ·

137. R. Washington and M. C. Todd, "Atmospheric controls on mineral dust emission from the Bodélé Depression, Chad: the role of the low level jet," Geophysical Research Letters, vol. 32, article 17701, 2005.

138. R. Washington, M. C. Todd, S. Engelstaedter, S. Mbainayel, and F. Mitchell, "Dust and the low-level circulation over the Bodélé Depression, Chad: observations from BoDEx 2005," Journal of Geophysical Research D, vol. 111, no. 3, Article ID D03201, 2006. · ·

139. R. Washington and M. C. Todd, "Atmospheric controls on mineral dust emission from the Bodélé Depression, Chad: the role of the low level jet," Geophysical Research Letters, vol. 32, no. 17, Article ID L17701, pp. 1–5, 2005. ·

140. M. C. Todd, R. Washington, S. Raghavan, G. Lizcano, and P. Knippertz, "Regional model simulations of the Bodélé low-level jet of Northern Chad during the Bodélé dust experiment (BoDEx 2005),"Journal of Climate, vol. 21, no. 5, pp. 995–1012, 2008. · ·

141. R. A. Tomas and P. J. Webster, "The role of inertial instability in determining the location and strength of near-equatorial convection," Quarterly Journal of the Royal Meteorological Society, vol. 123, no. 542, pp. 1445–1482, 1997.

142. S. A. Grodsky, J. A. Carton, and S. Nigam, "Near surface westerly wind jet in the Atlantic ITCZ,"Geophysical Research Letters, vol. 30, no. 19, pp. 1–4, 2003.

143. B. Pu and K. H. Cook, "Dynamics of the West African westerly jet," Journal of Climate, vol. 23, no. 23, pp. 6263–6276, 2010.

144. J. P. Grist, S. E. Nicholson, and A. I. Barcilon, "Easterly waves over Africa. Part II: observed and modeled contrasts between wet and dry years," Monthly Weather Review, vol. 130, no. 2, pp. 212–225, 2002.

145. B. Pu and K. H. Cook, "Role of the West African Westerly Jet in Sahel rainfall variations," Journal of Climate, vol. 25, pp. 2880–2896, 2012.

146. K. Abdou, D. J. Parker, B. Brooks, N. Kalthoff, and T. Lebel, "The diurnal cycle of lower boundary-layer wind in the West African Monsoon," Quarterly Journal of the Royal Meteorological Society, vol. 136, no. 1, pp. 66–76, 2010.

147. M. Lothon, F. Saïd, F. Lohou, and B. Campistron, "Observation of the diurnal cycle in the low troposphere of West Africa," Monthly Weather Review, vol. 136, no. 9, pp. 3477–3500, 2008.

148. B. Sultan, S. Janicot, and P. Drobinski, "Characterization of the diurnal cycle of the West African monsoon around the monsoon onset," Journal of Climate, vol. 20, no. 15, pp. 4014–4032, 2007.

149. C. L. Bain, D. J. Parker, C. M. Taylor, L. Kergoat, and F. Guichard, "Observations of the nocturnal boundary layer associated with the West African monsoon," Monthly Weather Review, vol. 138, no. 8, pp. 3142–3156, 2010.

150. P. Peyrille and J. Lafore, "An idealized two-dimensional framework to study the West African monsoon. Part II: large-scale advection and the diurnal cycle," Journal of the Atmospheric Sciences, vol. 64, no. 8, pp. 2783–2803, 2007.

151. J. M. Schrage and A. H. Fink, "Nocturnal continental low-level stratus over tropical West Africa: observations and possible mechanisms controlling its onset," Monthly Weather Review, vol. 140, pp. 1794–1809, 2012.

152. C. Lavaysse, C. Flamant, S. Janicot, and P. Knippertz, "Links between African easterly waves, midlatitude circulation and intraseasonal pulsations of the West African heat low," Quarterly Journal of the Royal Meteorological Society, vol. 136, no. 1, pp. 141–158, 2010. · ·

153. M. Biasutti, A. H. Sobel, and S. J. Camargo, "The role of the Sahara low in summertime Sahel rainfall variability and change in the CMIP3 models," Journal of Climate, vol. 22, no. 21, pp. 5755–5771, 2009. · ·

154. C. Lavaysse, C. Flamant, S. Janicot et al., "Seasonal evolution of the West African heat low: a climatological perspective," Climate Dynamics, vol. 33, no. 2-3, pp. 313–330, 2009. · ·

155. A. Mekonnen, C. D. Thorncroft, and A. R. Aiyyer, "Analysis of convection and its association with African easterly waves," Journal of Climate, vol. 19, no. 20, pp. 5405–5421, 2006. · ·

156. L. Bounoua and T. N. Krishnamurti, "Thermodynamic budget of the five day wave over the Saharan desert during summer," Meteorology and Atmospheric Physics, vol. 47, no. 1, pp. 1–25, 1991. · ·

157. C. Lavaysse, J. P. Chaboureau, and C. Flamant, "Dust impact on the west african heat low in summertime," Quarterly Journal of the Royal Meteorological Society, vol. 137, no. 658, pp. 1227–1240, 2011. · ·

158. C. M. Grams, S. C. Jones, J. H. Marsham, D. J. Parker, J. M. Haywood, and V. Heuveline, "The atlantic inflow to the saharan heat low: observations and modelling," Quarterly Journal of the Royal Meteorological Society, vol. 136, no. 1, pp. 125–140, 2010. · ·

159. C. Messager, D. J. Parker, O. Reitebuch, A. Agusti-Panareda, C. M. Taylor, and J. Cuesta, "Structure and dynamics of the Saharan atmospheric boundary layer during the West African monsoon onset: observations and analyses from the research flights of 14 and 17 July 2006," Quarterly Journal of the Royal Meteorological Society, vol. 136, no. 1, pp. 107–124, 2010. · ·

160. J. M. Prospero, P. Ginoux, O. Torres, S. E. Nicholson, and T. E. Gill, "Environmental characterization of global sources of atmospheric soil dust identified with the Nimbus 7 Total Ozone Mapping Spectrometer (TOMS) absorbing aerosol product," Reviews of Geophysics, vol. 40, no. 1, pp. 1–31, 2002.

161. J. P. Dunion and C. S. Velden, "The impact of the Saharan Air Layer on Atlantic tropical cyclone activity," Bulletin of the American Meteorological Society, vol. 85, no. 3, pp. 353–365, 2004. · ·

162. D. J. Parker, C. D. Thorncroft, R. R. Burton, and A. Diongue-Niang, "Analysis of the African easterly jet, using aircraft observations from the JET2000 experiment," Quarterly Journal of the Royal Meteorological Society, vol. 131, no. 608, pp. 1461–1482, 2005. · ·

163. H. F. Diaz, T. N. Carlson, and J. M. Propsero, "A study of the structure and dynamics of the Saharan air layer over the northern equatorial Atlantic during BOMEX," NOAA Technical Memorandum ERL WMPO-32, 1976.

164. S. Wong, A. E. Dessler, N. M. Mahowald, P. Yang, and Q. Feng, "Maintenance of lower tropospheric temperature inversion in the Saharan air layer by dust and dry anomaly," Journal of Climate, vol. 22, no. 19, pp. 5149–5162, 2009. · ·

165. V. M. Karyampudi, S. P. Palm, J. A. Reagen et al., "Validation of the Saharan dust plume conceptual model using lidar, Meteosat, and ECMWF data," Bulletin of the American Meteorological Society, vol. 80, no. 6, pp. 1045–1075, 1999.

166. F. Couvreux, C. Rio, F. Guichard, et al., "Initiation of daytime local convection in a semi-arid region analysed with high-resolution simulations and AMMA observations," Quarterly Journal of the Royal Meteorological Society, vol. 138, pp. 56–71, 2012.

167. S. W. Nesbitt, R. Cifelli, and S. A. Rutledge, "Storm morphology and rainfall characteristics of TRMM precipitation features," Monthly Weather Review, vol. 134, no. 10, pp. 2702–2721, 2006. · ·

168. C. Schumacher and R. A. Houze Jr., "Stratiform rain in the tropics as seen by the TRMM precipitation radar," Journal of Climate, vol. 16, no. 11, pp. 1739–1756, 2003.

169. B. Jackson, S. E. Nicholson, and D. Klotter, "Mesoscale convective systems over western equatorial Africa and their relationship to large-scale circulation," Monthly Weather Review, vol. 137, no. 4, pp. 1272–1294, 2009. · ·

170. L. Ricciardulli and P. D. Sardeshmukh, "Local time- and space scales of organized tropical deep convection," Journal of Climate, vol. 15, no. 19, pp. 2775–2790, 2002.

171. A. Gaye, A. Viltard, and P. de Félice, "Squall lines and rainfall over Western Africa during summer 1986 and 87," Meteorology and Atmospheric Physics, vol. 90, no. 3-4, pp. 215–224, 2005. · ·

172. R. N. Ferreira, T. Rickenbach, N. Guy, and E. Williams, "Radar observations of convective system variability in relationship to african easterly waves during the 2006 AMMA special observing period,"Monthly Weather Review, vol. 137, no. 12, pp. 4136–4150, 2009. · ·

173. A. H. Fink and A. Reiner, "Spatiotemporal variability of the relation between African Easterly Waves and West African Squall Lines in 1998 and 1999," Journal of Geophysical Research D, vol. 108, no. 11, article 4332, pp. 1–17, 2003.

174. A. G. Laing, R. Carbone, V. Levizzani, and J. Tuttle, "The propagation and diurnal cycles of deep convection in northern tropical Africa," Quarterly Journal of the Royal Meteorological Society, vol. 134, no. 630, pp. 93–109, 2008. · ·

175. K. I. Mohr and C. D. Thorncroft, "Intense convective systems in West Africa and their relationship to the African easterly jet," Quarterly Journal of the Royal Meteorological Society, vol. 132, no. 614, pp. 163–176, 2006. · ·

176. S. D. Nicholls and K. I. Mohr, "An analysis of the environments of intense convective systems in West Africa in 2003," Monthly Weather Review, vol. 138, pp. 3721–3739, 2010.

177. P. Knippertz and J. E. Martin, "Tropical plumes and extreme precipitation in subtropical and tropical West Africa," Quarterly Journal of the Royal Meteorological Society, vol. 131, no. 610, pp. 2337–2365, 2005. · ·

178. P. Knippertz, A. H. Fink, A. Reiner, and P. Speth, "Three late summer/early autumn cases of tropical-extratropical interactions

causing precipitation in Northwest Africa," Monthly Weather Review, vol. 131, no. 1, pp. 116–135, 2003.

179. P. Knippertz and J. E. Martin, "The role of dynamic and diabatic processes in the generation of cut-off lows over Northwest Africa," Meteorology and Atmospheric Physics, vol. 96, no. 1-2, pp. 3–19, 2007. · ·

180. F. Meier and P. Knippertz, "Dynamics and predictability of a heavy dry-season precipitation event over West Africa— sensitivity experiments with a global model," Monthly Weather Review, vol. 137, no. 1, pp. 189–206, 2009. · ·

181. K. Schepanski and P. Knippertz, "Soudano-Saharan depressions and their importance for precipitation and dust: a new perspective on a classical synoptic concept," Journal of Climate, vol. 137, pp. 1431–1445, 2011.

182. P. Knippertz and A. H. Fink, "Dry-season precipitation in tropical West Africa and its relation to forcing from the extratropics," Monthly Weather Review, vol. 136, no. 9, pp. 3579–3596, 2008. · ·

183. M. Geb, "Factors favouring precipitation in North Africa: seen from the viewpoint of present-day climatology," Global and Planetary Change, vol. 26, no. 1–3, pp. 85–96, 2000. · ·

184. J. Zawislak and E. J. Zipser, "Observations of seven African easterly waves in the east Atlantic during 2006," Journal of the Atmospheric Sciences, vol. 67, no. 1, pp. 26–43, 2010. · ·

185. A. Diedhiou, S. Janicot, A. Viltard, P. De Felice, and H. Laurent, "Easterly wave regimes and associated convection over West Africa and tropical Atlantic: results from the NCEP/NCAR and ECMWF reanalyses," Climate Dynamics, vol. 15, no. 11, pp. 795–822, 1999. · ·

186. I. Pytharoulis and C. Thorncroft, "The low-level structure of African easterly waves in 1995," Monthly Weather Review, vol. 127, no. 10, pp. 2266–2280, 1999.

187. A. H. Fink, D. G. Vincent, P. M. Reiner, and P. Speth, "Mean state and wave disturbances during Phases I, II and III of GATE based on ERA-40," Monthly Weather Review, vol. 132, pp. 1661–1683, 2004.

188. C. Thorncroft and K. Hodges, "African easterly wave variability and its relationship to Atlantic tropical cyclone activity," Journal of Climate, vol. 14, no. 6, pp. 1166–1179, 2001.

189. T. C. Chen, "Characteristics of African easterly waves depicted by ECMWF reanalyses for 1991–2000," Monthly Weather Review, vol. 134, no. 12, pp. 3539–3566, 2006. · ·

190. L. M. Druyan, M. Fulakeza, and P. Lonergan, "Mesoscale analyses of West African summer climate: focus on wave disturbances," Climate Dynamics, vol. 27, no. 5, pp. 459–481, 2006. · ·

191. R. S. Ross and T. N. Krishnamurti, "Low-level African easterly wave activity and its relation to Atlantic tropical cyclogenesis in 2001," Monthly Weather Review, vol. 135, no. 12, pp. 3950–3964, 2007. ·

192. G. J. Berry and C. Thorncroft, "Case study of an intense African easterly wave," Monthly Weather Review, vol. 133, no. 4, pp. 752–766, 2005. · ·

193. S. Leroux and N. M. J. Hall, "On the relationship between African easterly waves and the African easterly jet," Journal of the Atmospheric Sciences, vol. 66, no. 8, pp. 2303–2316, 2009. · ·

194. A. J. Matthews, "Intraseasonal variability over tropical Africa during northern summer," Journal of Climate, vol. 17, pp. 2427–2440, 2004.

195. G. J. Alaka and E. D. Maloney, "The influence of the MJO on upstream precursors to African Easterly Waves," Journal of Climate, vol. 25, pp. 3219–3236, 2012.

196. F. Mounier, G. N. Kiladis, and S. Janicot, "Analysis of the dominant mode of convectively coupled Kelvin waves in the West African monsoon," Journal of Climate, vol. 20, no. 8, pp. 1487–1503, 2007. · ·

197. A. Mekonnen, C. D. Thorncroft, A. R. Aiyyer, and G. N. Kiladis, "Convectively coupled Kelvin waves over tropical Africa during the boreal summer: structure and variability," Journal of Climate, vol. 21, no. 24, pp. 6649–6667, 2008. · ·

198. J. P. Grist, "Easterly waves over Africa. Part I: the seasonal cycle and contrasts between wet and dry years," Monthly Weather Review, vol. 130, no. 2, pp. 197–211, 2002.

199. E. H. Taleb and L. M. Druyan, "Relationships between rainfall and West African wave disturbances in station observations," International Journal of Climatology, vol. 23, no. 3, pp. 305–313, 2003. · ·

200. G. N. Kiladis, C. D. Thorncroft, and N. M. J. Hall, "Three-dimensional structure and dynamics of African easterly waves. Part I: observations," Journal of the Atmospheric Sciences, vol. 63, no. 9, pp. 2212–2230, 2006. · ·

201. C. L. Bain, D. J. Parker, N. Dixon et al., "Anatomy of an observed African easterly wave in July 2006,"Quarterly Journal of the Royal Meteorological Society, vol. 137, no. 657, pp. 923–933, 2011. · ·

202. D. Monkam, "The 6–9 day wave and rainfall modulation in northern Africa during summer 1981,"Journal of Geophysical Research D, vol. 108, no. 17, pp. 5–12, 2003.

203. P. J. Lamb, "Case studies of tropical Atlantic surface circulation patterns during recent sub-Saharan weather anomalies: 1967 and 1968," Monthly Weather Review, vol. 106, pp. 482–491, 1978.

204. P. J. Lamb, "Large-scale tropical Atlantic surface circulation patterns associated with sub-Saharan weather anomalies," Tellus, vol. 30, pp. 482–491, 1978.

205. C. K. Folland, T. N. Palmer, and D. E. Parker, "Sahel rainfall and worldwide sea temperatures, 1901–1985," Journal of Forecasting, vol. 1, pp. 21–56, 1986.

206. A. Giannini, R. Saravanan, and P. Chang, "Oceanic forcing of sahel rainfall on interannual to interdecadal time scales," Science, vol. 302, no. 5647, pp. 1027–1030, 2003. · ·

207. A. Giannini, R. Saravanan, and P. Chang, "Dynamics of the boreal summer African monsoon in the NSIPP1 atmospheric model," Climate Dynamics, vol. 25, pp. 517–535, 2005.

208. A. Giannini, M. Biasutti, and M. M. Verstraete, "A climate model-based review of drought in the Sahel: desertification, the re-greening and climate change," Global and Planetary Change, vol. 64, no. 3-4, pp. 119–128, 2008. · ·

209. D. P. Rowell, "The impact of Mediterranean SSTs on the Sahelian rainfall season," Journal of Climate, vol. 16, no. 5, pp. 849–862, 2003.

210. K. M. Lau, S. S. P. Shen, K. M. Kim, and H. Wang, "A multimodel study of the twentieth-century simulations of Sahel drought from the 1970s to 1990s," Journal of Geophysical Research D, vol. 111, no. 7, Article ID D07111, 2006. · ·

211. M. N. Ward, "Diagnosis and short-lead time prediction of summer rainfall in tropical North Africa at interannual and multidecadal timescales," Journal of Climate, vol. 11, no. 12, pp. 3167–3191, 1998.

212. M. Joly, A. Voldoire, H. Douville, P. Terray, and J. F. Royer, "African monsoon teleconnections with tropical SSTs: validation and evolution in a set of IPCC4 simulations," Climate Dynamics, vol. 29, no. 1, pp. 1–20, 2007. · ·

213. M. Joly and A. Voldoire, "Role of the Gulf of Guinea in the interannual variability of the West African monsoon: what do we learn from CMIP3 coupled simulations?" International Journal of Climatology, vol. 30, no. 12, pp. 1843–1856, 2010. · ·

214. E. Mohino, B. Rodríguez-Fonseca, T. Losada, et al., "Changes in the interannual SST-forced signals on West African rainfall. AGCM intercomparison," Climate Dynamics, vol. 37, no. 9-10, pp. 1707–1725, 2011. ·

215. K. M. Lau, K. M. Kim, Y. C. Sud, and G. K. Walker, "A GCM study of the response of the atmospheric water cycle of West Africa and the Atlantic to Saharan dust radiative forcing," Annales Geophysicae, vol. 27, no. 10, pp. 4023–4037, 2009. · ·

216. A. Konare, A. S. Zakey, F. Solmon et al., "A regional climate modeling study of the effect of desert dust on the West African monsoon," Journal of Geophysical Research D, vol. 113, no. 12, Article ID D12206, 2008. · ·

217. M. Yoshioka, N. M. Mahowald, A. J. Conley et al., "Impact of desert dust radiative forcing on sahel precipitation: relative importance of dust compared to sea surface temperature variations, vegetation changes, and greenhouse gas warming," Journal of Climate, vol. 20, no. 8, pp. 1445–1467, 2007. · ·

218. B. Rodríguez-Fonseca, S. Janicot, E. Mohino et al., et al., "Interannual and decadal SST-forced responses of the West African monsoon," Atmospheric Science Letters, vol. 12, pp. 67–74, 2011.

219. L. M. Druyan, "Studies of 21st-century precipitation trends over West Africa," International Journal of Climatology, vol. 31, no. 10, pp. 1415–1424, 2011. ··

220. H. Paeth and A. Hense, "SST versus climate change signals in West Africa n rainfall: 20th-century variations and future projections," Climatic Change, vol. 65, no. 1-2, pp. 179–208, 2004. ··

221. C. M. Patricola and K. H. Cook, "Northern African climate at the end of the twenty-first century: an integrated application of regional and global climate models," Climate Dynamics, vol. 35, no. 1, pp. 193–212, 2010. ··

222. F. Hourdin, I. Musat, F. Guichard et al., "Amma-Model intercomparison project," Bulletin of the American Meteorological Society, vol. 91, no. 1, pp. 95–104, 2010. ··

223. L. M. Druyan, J. Feng, K. H. Cook et al., "The WAMME regional model intercomparison study,"Climate Dynamics, vol. 35, no. 1, pp. 175–192, 2010. ··

224. K. H. Cook, G. A. Meehl, and J. M. Arblaster, "Monsoon regimes and processes in CCSM4. Part II: african and American monsoon systems," Journal of Climate, vol. 24, pp. 2609–2621, 2012.

225. G. Nikulin, C. Jones, F. Giorgi, et al., "Precipitation climatology in an ensemble of CORDEX-Africa regional climate simulations," Journal of Climate, vol. 25, pp. 6057–6078, 2012.

226. P. M. Ruti, J. E. Williams, F. Hourdin et al., "The West African climate system: a review of the AMMA model inter-comparison initiatives," Atmospheric Science Letters, vol. 12, no. 1, pp. 116–122, 2011. ··

227. M. Hoerling, J. Hurrell, J. Eischeid, and A. Phillips, "Detection and attribution of twentieth-century northern and southern African rainfall change," Journal of Climate, vol. 19, no. 16, pp. 3989–4008, 2006. ··

228. T. Losada, B. Rodríguez-Fonseca, S. Janicot, S. Gervois, F. Chauvin, and P. Ruti, "A multi-model approach to the Atlantic Equatorial mode: impact on the West African monsoon," Climate Dynamics, vol. 35, no. 1, pp. 29–43, 2010. ··

229. J. Sun, H. Wang, and W. Yuan, "Linkage of the boreal spring antarctic oscillation to the west african summer monsoon," Journal of the Meteorological Society of Japan, vol. 88, no. 1, pp. 15–28, 2010. ··

230. I. Polo, A. Ullmann, P. Roucou, and B. Fontaine, "Weather regimes in the Euro-Atlantic and Mediterranean sector, and relationship with West African rainfall over the 1989–2008 period from a self-organizing maps approach," Journal of Climate, vol. 24, no. 13, pp. 3423–3432, 2011. · ·

231. R. Zhang and T. L. Delworth, "Impact of Atlantic multidecadal oscillations on India/Sahel rainfall and Atlantic hurricanes," Geophysical Research Letters, vol. 33, Article ID L17712, 2006. ·

232. T. L. Delworth, R. Zhang, and M. E. Mann, "Decadal to centennial variability of the Atlantic from observations and models. Ocean circulation: mechanisms and impacts," Geophysical Monograph Series, vol. 173, pp. 121–148, 2007.

233. M. F. Ting, Y. Kushnir, R. Seager, and C. H. Li, "Robust features ofg Atlantic multidecadal variability and its climate impacts," Geophysical Research Letters, vol. 38, Article ID L17705, 2011.

234. T. M. Shanahan, J. T. Overpeck, K. J. Anchukaitis et al., "Atlantic forcing of persistent drought in West Africa," Science, vol. 324, no. 5925, pp. 377–380, 2009. · ·

235. D. L. R. Hodson, R. T. Sutton, C. Cassou, N. Keenlyside, Y. Okumura, and T. Zhou, "Climate impacts of recent multidecadal changes in Atlantic Ocean sea surface temperature: a multimodel comparison," Climate Dynamics, vol. 34, no. 7, pp. 1041–1058, 2010. · ·

236. H. Paeth and P. Friederichs, "Seasonality and time scales in the relationship between global SST and African rainfall," Climate Dynamics, vol. 23, no. 7-8, pp. 815–837, 2004. · ·

237. D. P. Rowell, "Teleconnections between the tropical Pacific and the Sahel," Quarterly Journal of the Royal Meteorological Society, vol. 127, pp. 1683–1706, 2001.

238. C. F. Ropelewski and M. S. Halpert, "Global and regional scale precipitation patterns associated with the El-Nino Southern oscillation," Monthly Weather Review, vol. 115, pp. 1606–1626, 1987.

239. C. F. Ropelewski and M. S. Halpert, "Precipitation patterns associated with the high index phase of the southern oscillation," Journal of Climate, vol. 2, pp. 268–284, 1989.

240. S. E. Nicholson and J. Kim, "The relationship of the El Nino Southern oscillation to African rainfall,"International Journal of Climatology, vol. 17, no. 2, pp. 117–135, 1997.

241. J. Bader and M. Latif, "The impact of decadal-scale Indian Ocean sea surface temperature anomalies on Sahelian rainfall and the North Atlantic Oscillation," Geophysical Research Letters, vol. 30, no. 22, pp. 2169–2173, 2003.

242. F. Raicich, N. Pinardi, and A. Navarra, "Teleconnections between Indian monsoon and Sahel rainfall and the Mediterranean," International Journal of Climatology, vol. 23, no. 2, pp. 173–186, 2003. · ·

243. C. E. Chung and V. Ramanathan, "Weakening of north Indian SST gradients and the monsoon rainfall in India and the Sahel," Journal of Climate, vol. 19, no. 10, pp. 2036–2045, 2006. · ·

244. J. Bader and M. Latif, "The 1983 drought in the West Sahel: a case study," Climate Dynamics, vol. 36, no. 3, pp. 463–472, 2011. · ·

245. J. Lu and T. L. Delworth, "Oceanic forcing of the late 20th century Sahel drought," Geophysical Research Letters, vol. 32, Article ID L22706, 2005.

246. J. Lu, "The dynamics of the Indian Ocean sea surface temperature forcing of Sahel drought," Climate Dynamics, vol. 33, no. 4, pp. 445–460, 2009. · ·

247. I. Polo, B. Rodríguez-Fonseca, T. Losada, and J. García-Serrano, "Tropical atlantic variability modes (1979–2002)—part I: time-evolving SST modes related to West African rainfall," Journal of Climate, vol. 21, no. 24, pp. 6457–6475, 2008. · ·

248. T. Jung, L. Ferranti, and A. M. Tompkins, "Response to the summer of 2003 Mediterranean SST anomalies over Europe and Africa," Journal of Climate, vol. 19, no. 20, pp. 5439–5454, 2006. · ·

249. M. Gaetani, B. Fontaine, P. Roucou, and M. Baldi, "Influence of the Mediterranean Sea on the West African monsoon: intraseasonal variability in numerical simulations," Journal of Geophysical Research-Atmospheres, vol. 115, no. D24, 2010. ·

250. B. Fontaine, J. Garcia-Serrano, P. Roucou et al., "Impacts of warm and cold situations in the Mediterranean basins on the West African monsoon: observed connection patterns (1979–2006) and climate simulations," Climate Dynamics, vol. 35, no. 1, pp. 95–114, 2010. · ·

251. J. G. Charney, "The dynamics of deserts and droughts," Quarterly Journal of the Royal Meteorological Society, vol. 101, pp. 193–202, 1975.

252. D. Entekhabi, "Recent advances in land-atmosphere interaction research," Reviews of Geophysics, vol. 33, no. 2, pp. 995–1003, 1995.

253. A. R. Lare and S. E. Nicholson, "Contrasting conditions of surface water balance in wet years and dry years as a possible land surface-atmosphere feedback mechanism in the West African Sahel," Journal of Climate, vol. 7, no. 5, pp. 653–668, 1994.

254. N. Zeng, J. D. Neelin, K. M. Lau, and C. J. Tucker, "Enhancement of interdecadal climate variability in the Sahel by vegetation interaction," Science, vol. 286, no. 5444, pp. 1537–1540, 1999.

255. C. M. Taylor, E. F. Lambin, N. Stephenne, R. J. Harding, and R. L. H. Essery, "The influence of land use change on climate in the Sahel," Journal of Climate, vol. 15, no. 24, pp. 3615–3629, 2002.

256. S. E. Nicholson, C. J. Tucker, and M. B. Ba, "Desertification, drought, and surface vegetation: an example from the West African Sahel," Bulletin of the American Meteorological Society, vol. 79, no. 5, pp. 815–829, 1998.

257. S. D. Prince, E. Brown De Colstoun, and L. L. Kravitz, "Evidence from rain-use efficiencies does not indicate extensive Sahelian desertification," Global Change Biology, vol. 4, no. 4, pp. 359–374, 1998.

258. J. F. Reynolds and D. M. Stafford Smith, Global Desertification: Do Humans Cause Deserts?Dahlem University Press, Berlin, Germany, 2002.

259. S. D. Prince, "Spatial and temporal scales for detection of desertification," in Global Desertification: Do Humans Cause Deserts? J. F. Reynolds and D. M. Stafford Smith, Eds., pp. 23–40, Dahlem University Press, Berlin, Germany, 2002.

260. C. A. Alo and G. Wang, "Role of dynamic vegetation in regional climate predictions over Western Africa," Climate Dynamics, vol. 35, no. 5, pp. 907–922, 2010.

261. B. J. Abiodun, J. S. Pal, E. A. Afiesimama, W. J. Gutowski, and A. Adedoyin, "Simulation of West African monsoon using RegCM3 Part II: impacts of deforestation and desertification," Theoretical

and Applied Climatology, vol. 93, no. 3-4, pp. 245–261, 2008. · ·

262. X. Zeng, M. Barlage, C. Castro, and K. Fling, "Comparison of land-precipitation coupling strength using observations and models," Journal of Hydrometeorology, vol. 11, no. 4, pp. 979–994, 2010. · ·

263. C. M. Taylor, P. P. Harrisa, and D. J. Parkerb, "Impact of soil moisture on the development of a sahelian mesoscale convective system: a case-study from the AMMA special observing period,"Quarterly Journal of the Royal Meteorological Society, vol. 136, no. 1, pp. 456–470, 2010. · ·

264. M. Rietkerk, V. Brovkin, P. M. van Bodegom et al., "Local ecosystem feedbacks and critical transitions in the climate," Ecological Complexity, vol. 8, no. 3, pp. 223–228, 2011. · ·

265. L. M. Druyan, M. Fulakeza, and P. Lonergan, "Land surface influences on the West African summer monsoon: implications for synoptic disturbances," Meteorology and Atmospheric Physics, vol. 86, no. 3-4, pp. 261–273, 2004. · ·

266. A. L. Steiner, J. S. Pal, S. A. Rauscher et al., "Land surface coupling in regional climate simulations of the West African monsoon," Climate Dynamics, vol. 33, no. 6, pp. 869–892, 2009. · ·

267. B. J. J. M. van den Hurk and E. van Meijgaard, "Diagnosing land-atmosphere interaction from a regional climate model simulation over West Africa," Journal of Hydrometeorology, vol. 11, no. 2, pp. 467–481, 2010. · ·

268. D. B. Clark, C. M. Taylor, and A. J. Thorpe, "Feedback between the land surface and rainfall at convective length scales," Journal Hydrometeor, vol. 5, pp. 625–639, 2004.

269. C. M. Taylor, D. J. Parker, C. R. Lloyd, and C. D. Thorncroft, "Observations of synoptic-scale land surface variability and its coupling with the atmosphere," Quarterly Journal of the Royal Meteorological Society, vol. 131, no. 607, pp. 913–937, 2005. · ·

270. C. M. Taylor, A. Gounou, F. Guichard et al., "Frequency of sahelian storm initiation enhanced over mesoscale soil-moisture patterns," Nature Geoscience, vol. 4, no. 7, pp. 430–433, 2011. · ·

271. L. Gantner and N. Kalthoff, "Sensitivity of a modelled life cycle of a mesoscale convective system to soil conditions over West Africa," Quarterly Journal of the Royal Meteorological Society, vol. 136, no. 1, pp. 471–482, 2010. · ·

272. R. A. Anthes, "Enhancement of convective precipitation by mesoscale variations in vegetative covering in semiarid regions," Journal of Climate & Applied Meteorology, vol. 23, no. 4, pp. 541–554, 1984.

273. B. Adler, N. Kalthoff, and L. Gantner, "Initiation of deep convection caused by land-surface inhomogeneities in West Africa: a modelled case study," Meteorology and Atmospheric Physics, vol. 112, no. 1-2, pp. 15–27, 2011. · ·

274. B. Adler, N. Kalthoff, and L. Gantner, "The impact of soil moisture inhomogeneities on the modification of a mesoscale convective system: an idealised model study," Atmospheric Research, vol. 101, no. 1-2, pp. 354–372, 2011. · ·

275. K. I. Mohr, R. D. Baker, W. K. Tau, and J. S. Famiglietti, "The sensitivity of West African convective-line water budgets to land cover," Journal of Hydrology, vol. 4, pp. 62–76, 2003.

276. D. J. Parker, "A simple model of coupled synoptic waves in the land surface and atmosphere of the northern Sahel," Quarterly Journal of the Royal Meteorological Society, vol. 134, no. 637, pp. 2173–2184, 2008. · ·

277. J. Schwendike, N. Kalthoff, and M. Kohler, "The impact of mesoscale convective systems on the surface and boundary-layer structure in West Africa: case-studies from the AMMA campaign 2006,"Quarterly Journal of the Royal Meteorological Society, vol. 136, no. 648, pp. 566–582, 2010. · ·

278. C. M. Taylor, R. J. Harding, A. J. Thorpe, and P. Bessemoulin, "A mesoscale simulation of land surface heterogeneity from HAPEX-Sahel," Journal of Hydrology, vol. 188-189, no. 1–4, pp. 1040–1066, 1997. · ·

279. C. M. Taylor and T. Lebel, "Observational evidence of persistent convective-scale rainfall patterns,"Monthly Weather Review, vol. 126, no. 6, pp. 1597–1607, 1998.

280. J.-P. Goutorbe, T. Lebel, A. Tinga et al., et al., "Hapex-sahel—a large-scale study of land-atmosphere interactions in the semiarid

tropics," Annales Geophysicae-Atmospheres Hydrospheres and Space Sciences, vol. 12, pp. 53–64, 1994.

281. C. M. Taylor, "Intraseasonal land-atmospheric coupling in the West African monsoon," Journal of Climate, vol. 21, no. 24, pp. 6636–6648, 2008. · ·

282. C. M. Taylor and R. J. Ellis, "Satellite detection of soil moisture impacts on convection at the mesoscale," Geophysical Research Letters, vol. 33, no. 3, Article ID L03404, 2006. · ·

283. V. Klupfel, N. Kalthoff, L. Gantner, and C. M. Taylor, "Convergence zones and their impact on the initiation of a mesoscale convective system in West Africa," Quarterly Journal of the Royal Meteorological Society, vol. 138, pp. 950–963, 2012.

284. D. Lauwaet, N. P. M. Lipzig, and K. Ridder, "The effect of vegetation changes on precipitation and Mesoscale Convective Systems in the Sahel," Climate Dynamics, vol. 33, no. 4, pp. 521–534, 2009. · ·

285. D. Lauwaet, N. P. M. van Lipzig, N. Kalthoff, and K. de Ridder, "Impact of vegetation changes on a mesoscale convective system in West Africa," Meteorology and Atmospheric Physics, vol. 107, no. 3, pp. 109–122, 2010. · ·

286. M. A. Gaertner, M. Domínguez, and M. Garvert, "A modelling case-study of soil moisture-atmosphere coupling," Quarterly Journal of the Royal Meteorological Society, vol. 136, no. 1, pp. 483–495, 2010. · ·

287. W. Moufouma-Okia and D. P. Rowell, "Impact of soil moisture initialisation and lateral boundary conditions on regional climate model simulations of the West African Monsoon," Climate Dynamics, vol. 35, no. 1, pp. 213–229, 2010. · ·

288. P. Knippertz and M. C. Todd, "The central west Saharan dust hot spot and its relation to African easterly waves and extratropical disturbances," Journal of Geophysical Research D, vol. 115, no. 12, Article ID D12117, 2010. · ·

289. C. Jones, N. Mahowald, and C. Luo, "The role of easterly waves on African desert dust transport,"Journal of Climate, vol. 16, pp. 3617–3628, 2003.

290. G. N›Tchayi Mbourou, J. J. Bertrand, and S. E. Nicholson, "The diurnal and seasonal cycles of wind-borne dust over africa north

of the equator," Journal of Applied Meteorology, vol. 36, no. 7, pp. 868–882, 1997.

291. P. J. deMott, K. Sassen, M. R. Poellot, et al., "African dust aerosols as atmospheric ice nuclei,"Geophysical Research Letters, vol. 30, article 1732, 2003.

292. W. J. Hui, B. I. Cook, S. Ravi, J. D. Fuentes, and P. D›Odorico, "Dust-rainfall feedbacks in the West African Sahel," Water Resources Research, vol. 44, no. 5, Article ID W05202, 2008. · ·

293. A. Wiacek, T. Peter, and U. Lohmann, "The potential influence of Asian and African mineral dust on ice, mixed-phase and liquid water clouds," Atmospheric Chemistry and Physics, vol. 10, no. 18, pp. 8649–8667, 2010. · ·

294. J. Huang, C. Zhang, and J. M. Prospero, "African aerosol and large-scale precipitation variability over West Africa," Environmental Research Letters, vol. 4, no. 1, Article ID 015006, 2009. · ·

295. O. Reale, K. M. Lau, and A. da Silva, "Impact of an interactive aerosol on the African Easterly Jet in the NASA GEOS-5 Global Forecasting System," Weather and Forecasting, vol. 26, pp. 504–519, 2011.

296. F. Solomon, N. Elguindi, and M. Mallet, "Radiative and climatic effects of dust over West Africa, as simulated by a regional climate model," Climate Research, vol. 52, article 2012, 2012.

297. L. Klüser and T. Holzer-Popp, "Relationships between mineral dust and cloud properties in the West African Sahel," Atmospheric Chemistry and Physics, vol. 10, no. 14, pp. 6901–6915, 2010. · ·

298. C. M. Patricola and K. H. Cook, "Atmosphere/vegetation feedbacks: a mechanism for abrupt climate change over northern Africa," Journal of Geophysical Research D, vol. 113, no. 18, Article ID D18102, 2008. · ·

299. P. Kabat, M. Claussen, P. A. Dirmeyer, et al., Vegetation, Water, Humans and the Climate: A New Perspective on an Interactive System, Springer, Berlin, Germany, 2004.

300. P. Demenocal, J. Ortiz, T. Guilderson et al., "Abrupt onset and termination of the African Humid Period: rapid climate responses to gradual insolation forcing," Quaternary Science Reviews, vol. 19, no. 1–5, pp. 347–361, 2000. · ·

301. S. Kroepelin, D. Verschuren, A.-M. Lézine, et al., "Climate-driven ecosystem succession in the Sahara: the past 6000 years," Science, vol. 320, pp. 765–768, 2008.

302. M. Claussen, "Modeling bio-geophysical feedback in the African and Indian monsoon region,"Climate Dynamics, vol. 13, no. 4, pp. 247–257, 1997.

303. Z. Liu, Y. Wang, R. Gallimore et al., "Simulating the transient evolution and abrupt change of Northern Africa atmosphere-ocean-terrestrial ecosystem in the Holocene," Quaternary Science Reviews, vol. 26, no. 13-14, pp. 1818–1837, 2007. · ·

304. M. Rietkerk, S. C. Dekker, P. C. De Ruiter, and J. Van De Koppel, "Self-organized patchiness and catastrophic shifts in ecosystems," Science, vol. 305, no. 5692, pp. 1926–1929, 2004. · ·

305. O. Lejeune, M. Tlidi, and P. Couteron, "Localized vegetation patches: a self-organized response to resource scarcity," Physical Review E, vol. 66, no. 1, Article ID 010901, 2002. · ·

306. M. Rietkerk and J. Van De Koppel, "Alternate stable states and threshold effects in semi-arid grazing systems," Oikos, vol. 79, no. 1, pp. 69–76, 1997.

307. R. H. H. Janssen, M. B. J. Meinders, E. H. van Nes, and M. Scheffer, "Microscale vegetation-soil feedback boosts hysteresis in a regional vegetation-climate system," Global Change Biology, vol. 14, no. 5, pp. 1104–1112, 2008. · ·

308. R. D. Koster, P. A. Dirmeyer, Z. Guo et al., "Regions of strong coupling between soil moisture and precipitation," Science, vol. 305, no. 5687, pp. 1138–1140, 2004. · ·

309. M. Scheffer, M. Holmgren, V. Brovkin, and M. Claussen, "Synergy between small—and large-scale feedbacks of vegetation on the water cycle," Global Change Biology, vol. 11, no. 7, pp. 1003–1012, 2005. · ·

310. P. D›Odorico, K. Caylor, G. S. Okin, and T. M. Scanlon, "On soil moisture-vegetation feedbacks and their possible effects on the dynamics of dryland ecosystems," Journal of Geophysical Research G, vol. 112, no. 4, Article ID G04010, 2007. · ·

311. S. C. Dekker, M. Rietkerk, and M. F. P. Bierkens, "Coupling microscale vegetation-soil water and macroscale vegetation-precipitation feedbacks in semiarid ecosystems," Global Change Biology, vol. 13, no. 3, pp. 671–678, 2007. · ·

312. S. C. Dekker, H. J. De Boer, V. Brovkin, K. Fraedrich, M. J. Wassen, and M. Rietkerk, "Biogeophysical feedbacks trigger shifts in the modelled vegetation-atmosphere system at multiple scales,"Biogeosciences, vol. 7, no. 4, pp. 1237–1245, 2010.

Chapter 4

Interannual and Intraseasonal Variability in Fine Mode Particles over Delhi: Influence of Meteorology

S. Tiwari[1], D. S. Bisht[1], A. K. Srivastava[1], G. P. Shivashankara[2], and R. Kumar[3]

[1]Indian Institute of Tropical Meteorology (Branch), Prof Ramnath Vij Marg, New Delhi 110060, India

[2]Department of Environmental Engineering, P. E. S. College of Engineering, Mandya, Karnataka 571401, India

[3]Sharda University, Knowledge Park III, Greater Noida 201306, India

ABSTRACT

Fine mode particles (i.e., $PM_{2.5}$) were collected at Delhi, India, for three consecutive years from January 2007 to December 2009 and were statistically analyzed. Daily mean mass concentration of $PM_{2.5}$ was found to be $108.81 \pm 75.5\ \mu g\ m^{-3}$ ranged from 12 to $367.9\ \mu g\ m^{-3}$, which is substantially higher than the Indian National Ambient Air Quality Standards (NAAQS). Among the measurements, ~69% of $PM_{2.5}$ samples exceeded 24 h Indian NAAQS of $PM_{2.5}$ level ($\mu g\,m^{-3}$); however, ~85% samples exceeded its annual level ($40\,\mu g\,m^{-3}$). Approximately 30% of PM2.5 mass was in the range of $40–80\,\mu g\,m^{-3}$, indicating abundance of fine particles over Delhi. Intraseasonal variability of $PM_{2.5}$ indicates highest mass concentration during postmonsoon ($154.31 \pm 81.62\,\mu g\,m^{-3}$), followed by winter ($150.81 \pm 74.65\,\mu g\,m^{-3}$), summer ($70.86 \pm 29.31\,\mu g\,m^{-3}$), and monsoon ($45.06 \pm 18.40\,\mu g\,m^{-3}$). In interannual variability, it was seen that in 2008, the fine mode particle was ~23% and ~36% higher as compared to 2007 and 2009, respectively. Significantly negative correlation was found between $PM_{2.5}$ and temperature (−0.59) as well as wind speed (−0.38). Higher concentration of $PM_{2.5}$ ($173.8\,\mu g\,m^{-3}$) was observed during calm conditions whereas low concentration ($79.18\,\mu g\,m^{-3}$) was observed when wind speed was >5 Km/hr. In winter, greater exposure risk is expected, as the pollutant often gets trapped in lower atmosphere due to stable atmospheric conditions.

INTRODUCTION

Aerosols, suspended in the atmosphere, are distributed through turbulence and direct atmospheric transport of air masses. These aerosols interact with Earth's energy budget, directly as well as indirectly. As a direct effect, aerosols scatter, absorb, and reflect solar energy that enters and exits in the Earth's atmosphere, while as an indirect effect, they altered the size, shape, and location of clouds and affected the precipitation in lower atmosphere [1–4]. Apart from this, atmospheric aerosols, especially fine particles, have received much attention during the last two decades due to their potential adverse impacts on human health and agricultural production. Particulate matters with aerodynamic diameters less than 2.5 μm (i.e., $PM_{2.5}$), called fine particles, have especially been found to be associated with increasing respiratory illness, carcinogens [5], asthma [6], and ultimately in increasing the number of premature deaths [7–9]. Many epidemic studies have linked airborne concentrations of $PM_{2.5}$ and PM_{10} with a variety of health problems, including morbidity and mortality [10].

Due to industrial and population growth, increased transportation system, burning of fossil fuel, high rate of urbanizations, and migrations are unavoidable in a developing country. India is the world's seventh largest country, second to China in its population. Rapid growth in megacities, especially in Delhi and Mumbai, is a cause of concern for air quality. Note that Delhi is the fourth most polluted and the seventh most populous metropolis in the world. According to a local survey, 30% of Delhi's population was found suffering from respiratory disorders due to air pollution, and this number is about 12 times the national average [11]. Over the station, the presence of industrial activity and traffic emissions are likely to be the most important sources of air pollutants. Increasing particulate matter has already noted to affect human health in megacities [12, 13]. As $PM_{2.5}$ particles have relatively large surface to volume ratio and longer residence times in the atmosphere, they possess a higher proportion of persistent organic compounds than larger particles [14]. In addition to this, high levels of $PM_{2.5}$ have been associated with amenity problems such as visibility degradation associated with haze [15].

The problem of pollution is not only the issue for Delhi, but also it is the case for several megacities of the entire world. Some studies on air quality assessment in Delhi have been carried out for PM emissions, effect of CNG regulations, air toxicity, and air quality index [16–18]. According to these studies, the transport sector of Delhi shares ~72% of the total airborne pollutants [16, 19]; however, major sources of air pollution in Delhi are emissions from vehicles (67%), coal based thermal power plants (13%), industrial units (12%), and domestic exhaust (8%). In 1991, the air pollutants daily emission loads over Delhi were ~1,450 metric tons (http://envfor.nic.in/divisions/cpoll/delpolln.html), which is still in the higher range exceeding prescribed standards of the World Health Organization [20]. The magnitude and urgency of the problem as a global environmental issue need a systematic understanding of the potential causes of pollution and their contribution to air quality. Mostly, in megacities, the main sources of fine particles are the combustion of fossil fuels from automobiles, construction equipments (mobile sources), furnaces, and power plants (stationary sources), where such particles are produced by combustion processes and mixed in the ambient air by mechanical processes [13, 21].

In the present study, fine mode particles (i.e., $PM_{2.5}$ mass concentrations) were obtained from Central Pollution Control Board (CPCB) during the period from January 2007 to December 2009 in Delhi. The observed aerosol data was analyzed and presented in the present study with objectives (i) to assess interannual and intraseasonal variability of $PM_{2.5}$ mass concentration over Delhi and its possible sources and (ii) to understand the effect of ambient meteorological parameters on aerosol formation and existence during different atmospheric conditions.

DATA COLLECTION AND SAMPLING SITE

Delhi, situated between 28°21'17" to 28°53' latitude and 76°20'37" to 77°20'37" longitude with height of ~218 meter above mean sea level, is around 160 km away in the south from the southern part of the Himalayas. It is bounded by the Thar desert of Rajasthan in the west, plains of central India in the south, and Indo-Gangetic Plains (IGP) in the east. The area of the city is 1,483 km^2 with ~18 million inhabitants. It experiences severe weather conditions between different seasons from hot and humid weather in summer to cold and dry weather in winter [22]. During the whole year, the prevailing wind was found to be easterly, northerly, and northwesterly, and to be strongest wind was in summer. Apart from such swings of weather in annual cycle, the whole northern part of India, especially the IGP, experiences thick foggy conditions during winter with lower boundary layer. More details about the station and the meteorological conditions are discussed elsewhere [23].

Hourly mass concentrations of $PM_{2.5}$ were collected for a period of three years from January 2007 to December 2009 which was monitored by CPCB (http://www.cpcb.nic.in/) at the Income Tax Office (ITO) intersection. CPCB is an independent governmental body which is responsible for monitoring the pollution levels at different environments across the country, including Delhi, under its National Ambient Air Quality Monitoring Network (NAAQMN). The sampling site is one of the highest traffic intersection zones, and a thermal power plant (747 MW) called "Indraprastha Thermal Power station" is located at about 500 m in the southeast azimuth. Further, the corresponding ambient meteorological parameters such as temperature, relative humidity (RH), and wind speed (WS) were collected from the India Meteorological Department (http://www.imd.gov.in/), Lodhi Road.

RESULTS AND DISCUSSION

Temporal Variability in Fine (PM$_{2.5}$) Mode Particle

Day to day variability in mass concentrations of fine mode particle (PM$_{2.5}$) during the study period from January 2007 to December 2009 was plotted and depicted in Figure 1. The daily mean mass concentration of PM$_{2.5}$ over Delhi during the study period was $108.81 \pm 75.5 \, \mu g \, m^{-3}$ that ranged from 12 to $367.9 \, \mu g \, m^{-3}$ which is substantially higher and far in excess of their annual averages stipulated by the Indian National Ambient Air Quality Standards (NAAQS; http://cpcb.nic.in/National_Ambient_Air_Quality_Standards.php; $40 \, \mu g \, m^{-3}$) and the US National Ambient Air Quality Standards (http://www.epa.gov/air/criteria.html; US NAAQS; $15 \, \mu gm^{-3}$). During the study period, ~69% PM$_{2.5}$ samples were found to exceed the 24 h limit of NAAQS for PM$_{2.5}$ standard ($60 \, \mu g \, m^{-3}$). Mass concentrations of PM$_{2.5}$ show considerable day to day variability, with the lowest value of $12 \, \mu g \, m^{-3}$ (on 18th August 2009) and the highest of $368 \, \mu g \, m^{-3}$ (on 1st March 2008), which could be due to the meteorological effect.

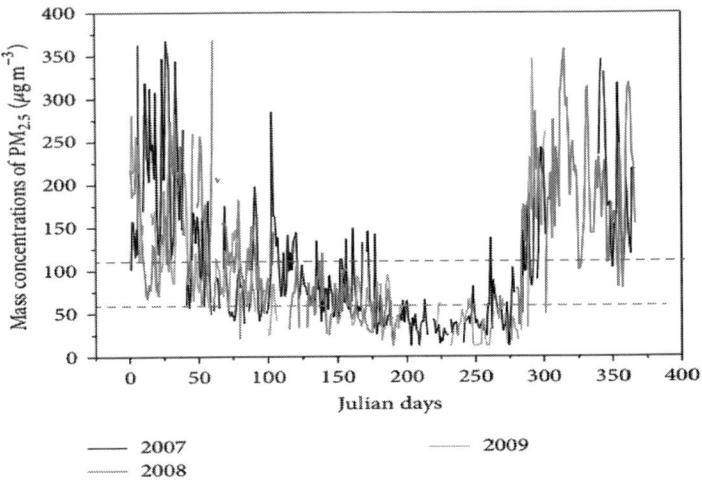

Figure 1: Day to day variability of mass concentrations of $PM_{2.5}$, during 2007 to 2009 (blue dash line—mean concentration of $PM_{2.5}$; red dash line—NAAQS).

Interannual monthly mean mass concentrations of $PM_{2.5}$ and its standard deviation during 2007 to 2009 were plotted and shown in Figure 2. Monthly mass concentrations of $PM_{2.5}$ were observed to be in the following order: January > February > October > April > March > June > May > September > July > August (2007), November > December > February > October > January > March > April > May > June > July (2008), and December > January > October > February > March > June > April > May > July > August > September (2009). On the basis of monthly analysis, $PM_{2.5}$ indicates the highest concentrations in winter and postmonsoon months. Also, we analyzed interannual variation of fine mode particles and were found the highest value during 2008 (135.44 ± 77 µg m⁻³, varying from 12.0 to 367.9 µg m⁻³) followed by 2007 (103.7 ± 77.5µg m⁻³, varying from 12.2 to 367.9 µg m⁻³) and 2009 (87.1 µg m⁻³, varying from 12.0 to 345.9 µg m⁻³). In 2008, the fine mode particle was higher by ~23% and 36% as compared to 2007 and 2009, respectively, which could be due to synoptic meteorological changes over the station (see Section 3.2). Intraseasonal variability of mass $PM_{2.5}$ was studied and shown in Figure 3. On the basis of annually season, the highest mass $PM_{2.5}$ concentrations were during postmonsoon (154.31 ± 81.62 µg m⁻³) followed by winter (150.81 ±

74.65 µg m⁻³), summer (**70.86 ± 29.31µg m⁻³**), and monsoon (45.06 ±
18.40µg m⁻³). In overall, the lower mass concentration of fine particle
was observed during monsoon due to washout effect, whereas the
higher concentration was during winter due to low level inversion.
$PM_{2.5}$ concentrations during monsoon in 2007 and 2009 were nearly
equal to the annual mean NAAQS except in 2008 (53.13 µg m⁻³);
however, it was approximately four times higher during postmonsoon
and winter whereas during summer, it was approximately double the
NAAQS.

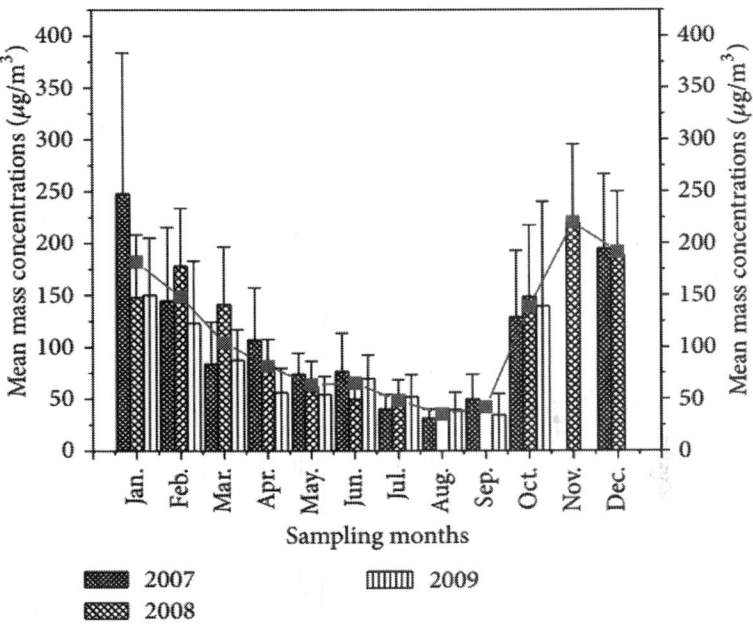

Figure 2: Yearwise monthly mean mass concentrations of $PM_{2.5}$, during 2007
to 2009.

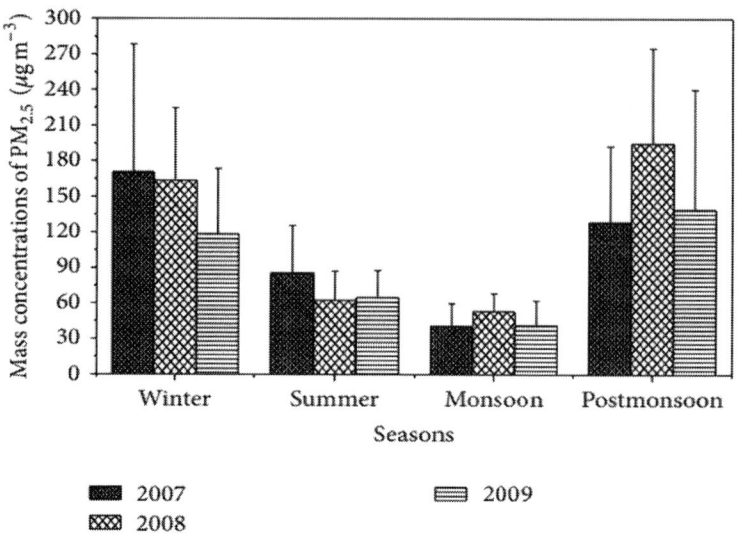

Figure 3: Seasonal variation of mass PM$_{2.5}$ concentrations during study period over Delhi.

Very high mass concentration of PM$_{2.5}$ during winter and postmonsoon is basically due to meteorological and emission effects [24, 25]. The mixing height is observed to be lower due to low temperature and calm winds during these periods. On the other hand, burning of fire crackers during Deepawali festival, which is generally celebrated in the last week of October or in the first week of November every year all over India, is another important cause of generation of particulate matters during postmonsoon season [26–29]. Minima were observed during monsoon season due to washout effect [24]. Bach et al. [30] reported an increase in total suspended particulate matters (TSPM) on an average by 300% at 14 locations; however, they also reported at one location that, due to fireworks on New Year's Eve, the lung penetrating size ranges particles increases up to 700%. Increase in particle number is witnessed in the accumulation mode range (>100 nm) during the Millennium Fireworks in Leipzig, Germany [31]. Further, Liu et al. [32] reported the chemical composition and particle size of typical firework mixtures. An enhancement in PM$_{2.5}$ up to 6 times and PM$_{10}$ up to 4 times on a lantern day (fireworks) in Beijing (China) is found relative to those over normal days [33].

Frequency distribution of $PM_{2.5}$ mass concentrations over Delhi during study period was also studied (Figure4). It is divided into nine different categories of 40 µg m⁻³ intervals within the limit of NAAQS from 0 to 360 µg m⁻³. In overall study, mass concentration of $PM_{2.5}$ skewed toward higher to lower concentrations in respect of corresponding spectrum except 0–40 µg m⁻³ (15%). The highest contribution was 30% (40–80 µg m⁻³) followed by 17% (80–120 µg m⁻³), 14% (120–160 µg m⁻³), 10% (160–200 µg m⁻³), 9% (200–240 µg m⁻³), 3% (240–280 µg m⁻³), 2% (280–320 µg m⁻³), and 1% (320–360 µg m⁻³). It clearly indicated that mass concentrations of $PM_{2.5}$ was higher (85%) than their National Ambient Air Quality Standard limits (40 µg m⁻³); however, ~30% of samples (significant fraction) were observed between 40 and 80 µg m⁻³ ranges, which indicates that the environment of Delhi is more dangerous in the health point of view due to high loadings of fine particles into the atmosphere of Delhi. This feature is basically due to emissions from anthropogenic sources and climatic conditions of Delhi.

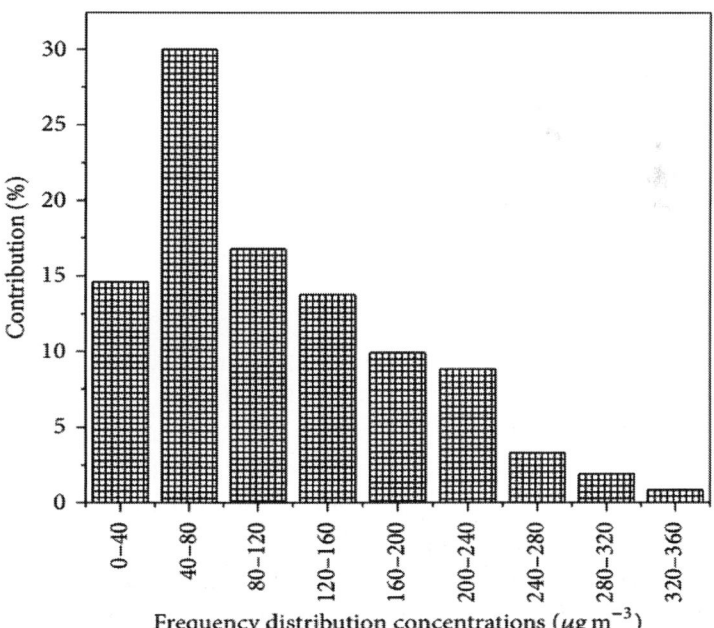

Figure 4: Frequency distribution of $PM_{2.5}$ over Delhi during study period.

Meteorological parameters such as temperature, relative humidity, wind speed, rainfall, and mixing height play crucial role in dispersion, transportation, and accumulation of the atmospheric pollutants. In general the atmosphere of Delhi during winter season is characterized by low relative humidity and low solar heating of land accompanied by low ventilation coefficients that result in less dispersion of aerosols, which in turn leads to an increase in the concentrations of fine mode particles. Due to this process, there is a greater exposure risk in trapping the air pollutants in the lower layer of the atmosphere thereby resulting in high mass concentrations of PM near to the surface. In such conditions, the probability of the formation of secondary aerosols is higher ([34] and references therein). Apart from this, long range transport of fine particles may also play crucial role during postmonsoon [24, 34]. Crop harvest and clearing agricultural land by burning the biomass during postmonsoon are common practices in the largely agricultural surroundings. The burning smoke reaches Delhi contributing to substantial smog formation and enhancement in PM and ozone levels [35]. In a recent study, Awasthi et al. [36] reported higher concentrations of $PM_{2.5}$ ($69\,\mu g\,m^{-3}$) as compared to NAAQS, which varied from 44 to $147\,\mu g\,m^{-3}$ at Patiala district of India, which is located northwest of Delhi near the foothills of the Himalayas. They found that the concentration of fine particulate matter was increased substantially (78%) during postmonsoon (October to November) due to burning of crop residue with the maxima observed from 100 to $147\,\mu g\,m^{-3}$ in 2009. Similar results with higher concentrations of aerosols are also reported by Badarinath et al. [37] using satellite-based measurements during exhaustive burning of rice crop residue in IGP region in the months of October to November. Therefore, the significantly high concentrations of fine size aerosols observed over Delhi have larger influence of certain anthropogenic activities related to agriculture during the winter season [24].

The daily mean mass concentration of PM ($108.81 \pm 75.5\mu g\,m^{-3}$) during the study period at Delhi is found to be relatively lower than the previously reported values ($171\,\mu g\,m^{-3}$) by Khillare et al. [38] during 2004, which can be attributed to introduction of metro-rail and use of CNG for public transport in Delhi. Lowering in the concentrations of $PM_{2.5}$ may increase in the proportion of petrol-fueled four- and two-wheeler vehicles, three-wheeler converted to CNG, and poor CNG kit in three-wheeler public transport system, which was first generation of CNG three-wheeler vehicles introduced in Delhi. Apart from these, the other significant fallout of the ruling was in the industrial sector—approximately 500 heavy industries were either shut down or relocated to areas outside the Delhi administrative boundaries. Yet, there remains a tremendous amount of potential to reduce the air pollution impacts in Delhi as the demand rises for infrastructure and services. Reynolds and Kandlikar [39] examined the opportunities for combined benefits of Delhi's fuel switching strategy—not only for local air pollution but also for climate related affects—and evaluated the potential for extending such services to other cities. Gadde et al. [40] have identified the contribution of ~14% (globally) of the open burnt rice straw in the states of Punjab, Haryana, and Uttar Pradesh. In addition to this, Delhi's low nighttime temperature during the winter (~2°C) and indoors and outdoors heating of biofuels for heating purpose are accomplished with small coal-burning boilers, stoves, and open burning of leaves and woods [41]. Guttikunda and Calori [42] have done a GIS-based emissions inventory at 1 km × 1 km spatial resolution for air pollution analysis in Delhi and found the mass concentrations of $PM_{2.5}$ (123 ± 87 $\mu g\,m^{-3}$) which is higher to present study. Dey et al. [43] have also reported higher outdoor fine particulate ($PM_{2.5}$) mass concentrations (148.4 ± 67$\mu g\,m^{-3}$) over Delhi. Overall, the level of $PM_{2.5}$ in Delhi is comparable to Beijing, China (115 $\mu g\,m^{-3}$) [44], and to the wintertime in California's San Joaquin Valley, USA (138 $\mu g\,m^{-3}$) [45], and is higher than the Belgrade City (75 $\mu g\,m^{-3}$) [46].

Recent increase in the proportion of diesel cars and diesel light trucks appears to be the cause of increase of PM concentrations. Besides this, the particulate pollutants in Delhi environment are also contributed to nonexhaust particles originating from wear and corrosion of road pavements, vehicle components, and particles originating in surroundings as well as industrial processes, increasing construction activities, loss of vegetation, and thermal power plants. The four wheelers, which were converted from petrol to compressed natural gas (CNG), are characterized by poor quality of piston rings as well as the improper maintenance of air filters, which generate white smoke, causing increase in PM levels [47]. Apart from generation of particles in the vicinity of the city, particles from nearby thermal power plants also contribute significantly to the particle level of Delhi. Note that three major power plants, namely, Badarpur, Indraprastha, and Rajghat with the total electricity generation capacity of 1,087 MWs, are situated in the vicinity of measurement location. These power plants produce nearly 6,000 metric tons (Badarpur 3,500–4,000, Indraprastha 1,200–1,500, and Rajghat 600–800) of fly ash per day and are responsible for as much as 10% of the total air pollution load. Furthermore, Delhi is also dealing with massive dust due to continuous constructional activities and a failed effort to control burning of garbage and biomass.

Effect of Meteorology on $PM_{2.5}$ Mass Concentrations

Yearly mean mass concentrations of $PM_{2.5}$, temperature (Temp.), relative humidity (RH), and wind speed (WS) along with their standard deviation, minimum, and maximum values in Delhi during 2007 to 2009 are assembled in Table 1. Large variations were seen in interannual and intraseasonal (as discussed in Section 3.1) mass concentrations of fine mode particles over Delhi during the study period, which is directly influenced by ambient meteorological conditions. The annual mean mass concentration of $PM_{2.5}$ during 2008 was higher than in 2007 and 2009, although very good agreement was seen between mass concentrations and meteorological parameters such as WS, temperature, and RH. The highest mean WS (5.2 Km/hr) was observed in 2008, whereas the lowest was in 2009 (4.7 Km/hr) and the intermediate was in 2007 (5.0 Km/hr). Seasonwise occurrence of calm conditions was studied and higher occurrence of calm conditions was observed during winter period of 2008 as compared to 2007 and 2009; however, there are less calm conditions observed in other seasons. The RH was found similar to WS with lower in 2009 (mean RH: 59%) as compared to 2007 (mean RH: 64%) and 2008 (mean RH: 63%). There is no significant relationship seen during interannual variability between mass concentrations with RH; however, good agreement was seen on intraseasonal variability on mass concentrations and RH. Seasonwise variations between them have been studied and higher concentration of fine particles was observed during winter period as in 2007 (171 $\mu g\,m^{-3}$), 2008 (163 $\mu g\,m^{-3}$), and 2009 (118 $\mu g\,m^{-3}$) and corresponding RH was 72%, 62%, and 64%, respectively. In another study, Singh [48] suggested that RH plays a very crucial role in altering the radiative properties of atmospheric aerosols as hygroscopic nature. A very good agreement between columnar aerosol optical depth and

RH over Rajkot was observed by Ranjan et al. [49]. Devara and Raj [50] observed that the higher relative humidity and lower temperature during monsoon period at Pune caused the growth of cloud droplets which results in higher rainfall. In the case of temperature, it was higher during 2009 (Mean: 26°C) compared to 2007 (Mean: 25.4°C) and 2008 (Mean: 25.1°C). Also, large variation was observed between maximum and minimum temperature during interannual variability and varied from 9.4 to 39.0°C in 2007, 9.2 to 35.0°C in 2008, and 11.8 to 37.8°C in 2009. Statistically, the variability in temperature was seen during intraseasonal with highest magnitude during summer (31.5°C) followed by monsoon (31.0°C), postmonsoon (23.5°C), and winter (18.0°C).

Table 1: Yearly mean mass concentrations of $PM_{2.5}$, temperature, relative humidity, and wind speed along with their standard deviation (std.) and minimum (min) and maximum (max) values in Delhi during 2007 to 2009

Year	Mass concentration of PM2.5 $\mu(g\,m{-}3)$			Temperature (°C)			Relative humidity (%)			Wind speed (km/hr)		
	2007	2008	2009	2007	2008	2009	2007	2008	2009	2007	2008	2009
Mean	77.5	135.4	87.1	25.4	25.1	26.0	64.2	63.9	59.1	5.0	5.2	4.7
Std.	88.8	77.8	62.2	7.0	6.5	6.9	45.9	15.2	15.9	3.4	3.5	3.9
Max	367.3	367.9	345.9	39.0	35.0	37.8	88.0	97.0	97.0	18.9	18.9	19.4
Min	12.2	24.7	12.0	9.4	9.2	11.8	24	25	20	Calm	Calm	Calm

Regression analysis between daily mass concentrations of $PM_{2.5}$ and meteorological parameters such as WS, temperature, and RH during study period was performed and depicted in Figure 5 (Figure 5(a): WS; Figure5(b): temperature, and Figure 5(c): RH). Very good agreement was observed between mass concentrations with WS and temperature; however, very week relationships were seen between PM and RH in the overall study due to large variability in RH during different seasons. A good agreement was seen during intraseasonal which was discussed earlier. Significant correlation between $PM_{2.5}$ and WS over Delhi was observed (−0.38 (r), slope: Y = 6.97725-0.01784 X; N=755; P<0.0001), which is significant at 99% confidence level (Figure 5(a)). The mass concentrations were separated in different WS (calm conditions and WS ≤ 5 Km/hr and ≥5 Km/hr). Very high mass concentrations (173.8 µg m⁻³) of PM were observed when WS was in calm conditions; however, when the WS was in <5 Km/hr, it was still showing higher mass concentration (124.77 µg m⁻³) which is three times higher than NAAQS. Very low PM mass concentrations (79.18 µg m⁻³) were observed in higher WS (>5 Km/hr). Also, it was seen that the concentrations of pollutants decrease effectively with increasing wind speed, suggesting the dilution of pollutants through dispersion. Similarly to WS, the correlation between PM and temperature was also observed to be very high (-0.58 (r)), which is also significant at 99% confidence level (slope: Y=31.21856-0.05543 X; N=755; P<0.0001) (Figure 5(b)). We also separated the concentrations against temperature (indicated as vertical line) and found that the higher concentrations (164.66 µg m⁻³) were observed when the temperature was less than 22.5°C; however, very low concentrations (77.14 µg m⁻³) were observed to be >22.5°C. Wallace and Kanaroglou [51] studied the relationship between temperature and fine particulate matter during day and night at Ontario, Canada, for the period from 2003 to 2007. They found that in nighttime, the 54% concentrations of fine particles increased due to low level inversion and lower temperature but it was found opposite during daytime and the concentration decreased about 14% during daytime. Due to large variability during day and night temperature along with interannual and intraseasonal variability, very poor relationship (0.14) between RH and mass concentrations of PM was observed during the study period. Interesting results were seen in the separation of lower (<55%) and higher (>55%) RH; in case of <55% RH, the lower mass concentration (96.96 µg m⁻³) was observed;

however, in >55% RH, the mass concentration was found to be relatively higher (114.6 µg m⁻³). High wind speed was associated with lower pollutant levels in most of the cases, whereas in the case of RH and temperature, it was opposite. On the basis of these relationships, it can be mentioned that the interplay of meteorological variables with pollution plays a key role in assessing the impact of pollution for a region. Seasonal correlation analysis between fine particle and meteorological parameters was also performed and depicted in Table 2. The WS and temperature was found to be negatively correlated during winter and post-monsoon seasons; however, a poor correlation was observed during summer and monsoon. Very interesting results were seen in the case of RH having positive relationship (=0.35) during postmonsoon whereas significant negative correlation (−0.48) was seen during summer. No relationship was observed during monsoon and winter. It is clearly indicated that meteorological parameters play a vital role in dispersion and accumulation of aerosols in different seasons. Chate and Devara [52] studied the impact of RH on nucleation mode particle during winter of 1997 and 1998 and found that they have positive impact on growth in submicron aerosols (0.013 to 0.133 µg m⁻³) during cold season; however, during the same period in the present study, we have not found any significant relation in larger particle (2.5 µg m⁻³) as compared to nucleation aerosols. In an other study, Cheng and Lam [53] investigated the impact of wind on TSP concentrations in Hong Kong and found a similar relationship. Also, Chaloulakou et al. [54] investigated the relationship among PM and meteorological parameters such as wind speed over Athens, Greece, and found very good agreement with wind speed (r = -0.43) and mass PM. They also found good agreement between temperature and fine particles below 10°C (−0.36) and found a positive correlation above 30°C (0.41). Further, $PM_{2.5}$ concentrations measured near a highly trafficked road in Paris were found to be inversely proportional to the wind speed [55].

Table 2: Interannual and intraseasonal correlation coefficient between $PM_{2.5}$, and meteorological parameters (wind speed: WS, temperature: Temp., and relative humidity: RH) in Delhi during 2007 to 2009

	2007	2008	2009	Overall
	PM2.5	PM2.5	PM2.5	PM2.5

WS	−0.30	−0.56	−0.30	−0.38
Temp.	−0.60	−0.54	−0.56	−0.59
RH	0.03	0.11	−0.01	0.14
	Winter	Summer	Monsoon	Postmonsoon
	PM2.5	PM2.5	PM2.5	PM2.5
WS	−0.50	−0.16	−0.09	−0.58
Temp.	−0.58	−0.02	−0.11	−0.51
RH	0.19	−0.48	−0.18	0.35

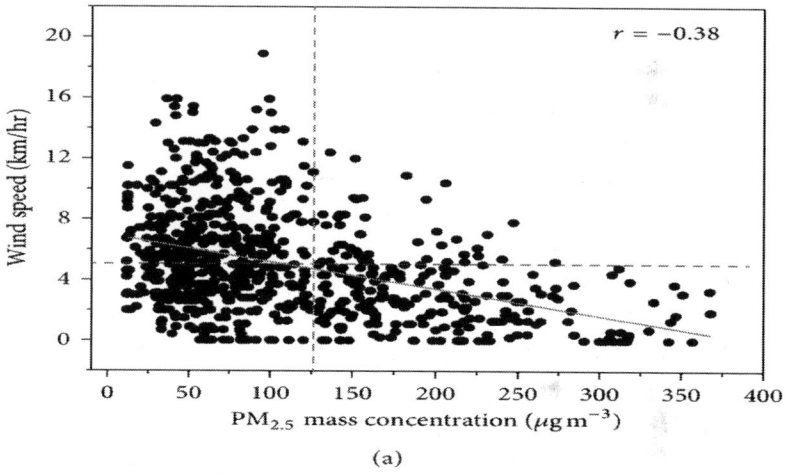

(a)

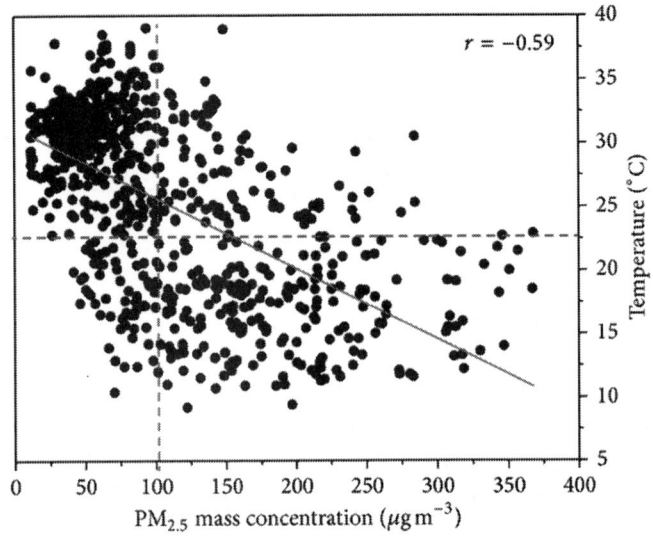

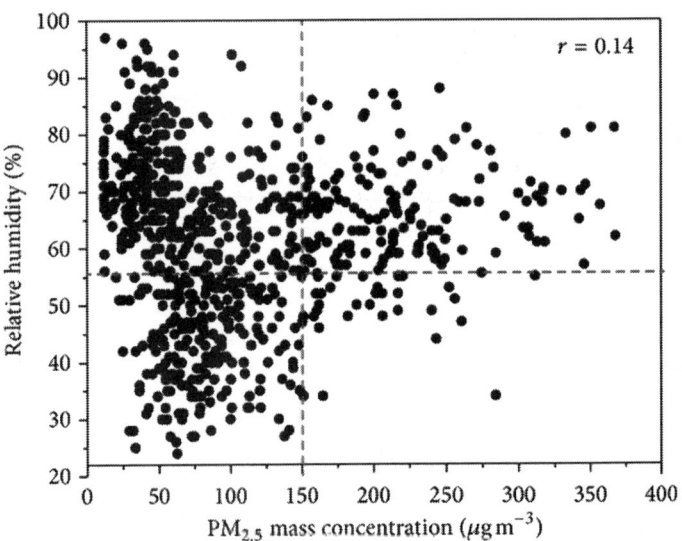

Figure 5: Day to day relation between wind speed and $PM_{2.5}$ (a), temperature and $PM_{2.5}$(b), and relative humidity and $PM_{2.5}$ (c) during 2007 to 2009. Vertical and horizontal lines indicate the dominance of the metrological parameters and $PM_{2.5}$ mass concentration.

CONCLUSIONS

Mass concentrations of $PM_{2.5}$ were collected during three consecutive years over Delhi at a busy traffic intersection at I.T.O., from January 2007 to December 2010. The data were analyzed for annual and seasonal variations of $PM_{2.5}$ mass concentration and their results are summarized and given below.

The capital of India, Delhi, is found to be heavily loaded with fine particulate matter ($PM_{2.5}$) showing daily mean mass concentration ($108.81 \pm 75.5 \mu g\,m^{-3}$) that ranged from 12 to $367.9\,\mu g\,m^{-3}$ which is substantially higher than the Indian NAAQS. Approximately 69% samples of $PM_{2.5}$ mass were exceeded to 24 h Indian NAAQS $PM_{2.5}$ level ($60\,\mu g\,m^{-3}$) whereas 85% samples were exceeded to its annual level ($40\,\mu g\,m^{-3}$). Most of the samples existed between 40 and $80\,\mu g\,m^{-3}$ (~30%) and indicated that the environment of Delhi is susceptible to a health point of view, generated due to anthropogenic emissions and meteorological conditions over Delhi. Intraseasonal variability of $PM_{2.5}$ indicates the highest mass concentrations during postmonsoon ($154.31 \pm 81.62 \mu g\,m^{-3}$) followed by winter ($150.81 \pm 74.65 \mu g\,m^{-3}$), summer ($70.86 \pm 29.31 \mu g\,m^{-3}$), and monsoon ($45.06 \pm 18.40 \mu g\,m^{-3}$). In 2008, the fine mode particle was ~23% and ~36% higher as compared to 2007 and 2009, respectively; these large interannual variations are found due to synoptic meteorological changes. Significant negative correlations are found between $PM_{2.5}$ and temperature (−0.59) as well as wind speed (−0.38). Higher concentrations of $PM_{2.5}$ ($173.8\,\mu g\,m^{-3}$) were observed during calm conditions, whereas $124.77\,\mu g\,m^{-3}$ was observed when WS was below 5 Km/hr. Very low PM mass concentrations ($79.18\,\mu g\,m^{-3}$) were observed in higher WS (>5 Km/hr). The high correlation is expected due to the cause that concentrations of pollutants decrease effectively with increasing WS and temperature, which suggests the dilution of pollutants into the atmosphere. In winter, greater exposure risk is expected, as the pollutant often gets trapped in lower atmosphere due to stable atmospheric conditions, thereby leading to higher levels. On the basis of the present study, it can be concluded that the interplay of meteorological variables with pollution plays crucial role in assessing the impact of pollution.

ACKNOWLEDGMENTS

The authors gratefully thank Professor B. N. Goswami, Director of the Indian Institute of Tropical Meteorology, Pune for their encouragement and support during the preparation of this paper. Authors also thank the Central Pollution Control Board and the India Meteorological Department for data generation.

REFERENCES

1. S. Dey, S. N. Tripathi, R. P. Singh, and B. Holben, "Influence of dust storms on the aerosol optical properties over the Indo-Gangetic basin," Journal of Geophysical Research D, vol. 109, no. 20, Article ID D20211, 13 pages, 2004.

2. S. Singh, S. Nath, R. Kohli, and R. Singh, "Aerosols over Delhi during pre-monsoon months: characteristics and effects on surface radiation forcing," Geophysical Research Letters, vol. 32, no. 13, Article ID L13808, 4 pages, 2005.

3. G. Pandithurai, S. Dipu, K. K. Dani et al., "Aerosol radiative forcing during dust events over New Delhi, India," Journal of Geophysical Research D, vol. 113, Article ID D13209, 13 pages, 2008.

4. K. A. Srivastava and S. N. Tripathi, "Numerical study for production of space charge within the stratiform cloud," Journal of Earth System Science, vol. 119, no. 5, pp. 627–638, 2010.

5. D. W. Dockery and C. A. Pope, "Acute respiratory effects of particulate air pollution," Annual Review of Public Health, vol. 15, pp. 107–132, 1994.

6. W. P. Anderson, C. M. Reid, and G. L. Jennings, "Pet ownership and risk factors for cardiovascular disease," Medical Journal of Australia, vol. 157, no. 5, pp. 298–301, 1992.

7. J. Chen, R. P. Wildman, D. Gu et al., "Prevalence of decreased kidney function in Chinese adults aged 35 to 74 years," Kidney International, vol. 68, no. 6, pp. 2837–2845, 2005.

8. F. Dominici, R. D. Peng, M. L. Bell et al., "Fine particulate air pollution and hospital admission for cardiovascular and respiratory

diseases," Journal of the American Medical Association, vol. 295, no. 10, pp. 1127–1134, 2006.

9.   G. E. R. Schwartz, L. G. S. Russek, L. A. Nelson, and C. Barentsen, "Accuracy and replicability of anomalous after-death communication across highly skilled mediums," Journal of the Society for Psychical Research, vol. 65, no. 862, pp. 1–25, 2001.

10. Y. Wang, C. Lee, S. Tiep et al., "Peroxisome-proliferator-activated receptor   activates fat metabolism to prevent obesity," Cell, vol. 113, no. 2, pp. 159–170, 2003.

11. M. Kandlikar, "The causes and consequences of particulate air pollution in urban India: a synthesis of the science," Annual Review of Energy and the Environment, vol. 25, pp. 629–684, 2000.

12. Central pollution Control Board (CPCB), 2008 Epidemiological study on effect of air pollution on human health (adults) in Delhi. Environmental Health Series: EHS/1/2008, http://www.cpcb.nic.in.

13. S. Madronich, "Chemical evolution of gaseous air pollutants down-wind of tropical megacities: Mexico City case study," Atmospheric Environment, vol. 40, no. 31, pp. 6012–6018, 2006.

14. R. Jaenicke, "Protein folding and Protein Association," Angewandte Chemie, vol. 23, no. 6, pp. 395–413, 1984.

15. J. W. Milne, D. B. Roberts, S. J. Walk, and D. J. William, "Sources of Sydney brown haze," in The Urban Atmosphere—Sydney. A Case Study, CSIRO, Highett, Australia, 1982.

16. P. Goyal and S. Sidhartha, "Present scenario of air quality in Delhi: a case study of CNG implementation," Atmospheric Environment, vol. 37, no. 38, pp. 5423–5431, 2003.

17. A. Kumar and T. C. Foster, "Shift in induction mechanisms underlies an age-dependent increase in DHPG-induced synaptic depression at CA3-CA1 synapses," Journal of Neurophysiology, vol. 98, no. 5, pp. 2729–2736, 2007.

18. B. Bishoi, A. Prakash, and V. K. Jain, "A comparative study of air quality index based on factor analysis and US-EPA methods for an urban environment," Aerosol and Air Quality Research, vol. 9, no. 1, pp. 1–17, 2009.

19. V. Kathuria, "Impact of CNG on Delhi's air pollution," Economic and Political Weekly, vol. 40, pp. 1907–1916, 2005.

20. B. R. Gurjar, J. A. Van Aardenne, J. Lelieveld, and M. Mohan, "Emission estimates and trends (1990–2000) for megacity Delhi and implications," Atmospheric Environment, vol. 38, no. 33, pp. 5663–5681, 2004.

21. A. Faiz, C. Weaver, K. Sinha, M. Walsh, and J. Carbajo, Air Pollution from Motor Vehicles: Issues and Options for Developing Countries, The World Bank, Washington, DC, USA, 1992.

22. A. K. Srivastava, S. Singh, S. Tiwari, V. P. Kanawade, and D. S. Bisht, "Variation between near-surface and columnar aerosol characteristics during the winter and summer at Delhi in the Indo-Gangetic Basin," Journal of Atmospheric and Solar-Terrestrial Physics, vol. 77, pp. 57–66, 2012.

23. S. Tiwari, A. K. Srivastava, D. S. Bisht, P. Parmita, M. K. Srivastava, and S. D. Attri, "Diurnal and seasonal variations of black carbon and $PM_{2.5}$ over New Delhi, India: influence of meteorology,"Atmospheric Research, vol. 125-126, pp. 50–62, 2013.

24. S. Tiwari, D. M. Chate, P. Pragya, K. Ali, and D. S. Bisht, "Variations in mass of the PM_{10}, $PM_{2.5}$ and PM_1 during the monsoon and the winter at New Delhi," Aerosol and Air Quality Research, vol. 12, no. 1, pp. 20–29, 2012.

25. S. Tiwari, A. K. Srivastava, D. S. Bisht, and P. D. Safai, "Assessment of carbonaceous aerosol over Delhi in the Indo-Gangetic Basin: characterization, sources and temporal variability," Natural Hazards, vol. 65, pp. 1745–1764, 2013.

26. S. Tiwari, D. M. Chate, M. K. Srivastava et al., "Statistical evaluation of PM_{10} and distribution of PM_1, $PM_{2.5}$, and PM_{10} in ambient air due to extreme fireworks episodes (Deepawali festivals) in megacity Delhi," Natural Hazards, vol. 61, no. 2, pp. 521–531, 2012.

27. A. K. Attri, U. Kumar, and V. K. Jain, "Formation of ozone by fireworks," Nature, vol. 411, no. 6841, pp. 1015–1021, 2001.

28. R. P. Singh, S. Dey, and B. Holben, "Aerosol behaviour in Kanpur during Diwali festival," Current Science, vol. 84, no. 10, pp. 1302–1303, 2003.

29. S. C. Barman, R. Singh, M. P. S. Negi, and S. K. Bhargava, "Fine particles ($PM_{2.5}$) in ambient air of Lucknow city due to fireworks on Diwali festival," Journal of Environmental Biology, vol. 30, no. 5, pp. 625–632, 2009.

30. W. Bach, A. Daniels, L. Dickinson et al., "Firework's pollution and health," International Journal of Environmental Studies, vol. 7, pp. 183–192, 1975.

31. B. Wehner, A. Wiedensohler, and J. Heintzenberg, "Submicrometer aerosol size distributions and mass concentration of the Millennium fireworks 2000 in Leipzig, Germany," Journal of Aerosol Science, vol. 31, no. 12, pp. 1489–1493, 2000.

32. D. Liu, D. Rutherford, M. Kinsey, and K. A. Prather, "Real-time monitoring of pyrotechnically derived aerosol particles in the troposphere," Analytical Chemistry, vol. 69, no. 10, pp. 1808–1814, 1997.

33. Y. Wang, G. Zhuang, C. Xu, and Z. An, "The air pollution caused by the burning of fireworks during the lantern festival in Beijing," Atmospheric Environment, vol. 41, no. 2, pp. 417–431, 2007.

34. A.-P. Hyvärinen, H. Lihavainen, M. Komppula et al., "Aerosol measurements at the Gual Pahari EUCAARI station: preliminary results from in-situ measurements," Atmospheric Chemistry and Physics, vol. 10, no. 15, pp. 7241–7252, 2010.

35. NASA (National Aeronautics and Space Administration), Top Science, Exploration and Discovery Stories of 2008.

36. A. Awasthi, R. Agarwal, S. K. Mittal, N. Singh, K. Singh, and P. K. Gupta, "Study of size and mass distribution of particulate matter due to crop residue burning with seasonal variation in rural area of Punjab, India," Journal of Environmental Monitoring, vol. 13, no. 4, pp. 1073–1081, 2011.

37. K. V. S. Badarinath, T. R. K. Chand, and V. K. Prasad, "Agriculture crop residue burning in the Indo-Gangetic Plains: a study using IRS-P6 AWiFS satellite data," Current Science, vol. 91, no. 8, pp. 1085–1089, 2006.

38. P. S. Khillare, T. Agarwal, and V. Shridhar, "Impact of CNG implementation on PAHs concentration in the ambient air of Delhi: a comparative assessment of pre- and post-CNG scenario,"

Environmental Monitoring and Assessment, vol. 147, no. 1–3, pp. 223–233, 2008.

39.   C. C. O. Reynolds and M. Kandlikar, "Climate impacts of air quality policy: switching to a natural gas-fueled public transportation system in New Delhi," Environmental Science and Technology, vol. 42, no. 16, pp. 5860–5865, 2008.

40.   B. Gadde, S. Bonnet, C. Menke, and S. Garivait, "Air pollutant emissions from rice straw open field burning in India, Thailand and the Philippines," Environmental Pollution, vol. 157, no. 5, pp. 1554–1558, 2009.

41.   K. Ali, G. A. Momin, S. Tiwari, P. D. Safai, D. M. Chate, and P. S. P. Rao, "Fog and precipitation chemistry at Delhi, North India," Atmospheric Environment, vol. 38, no. 25, pp. 4215–4222, 2004.

42.   S. K. Guttikunda and G. Calori, "A GIS based emissions inventory at 1 km × 1 km spatial resolution for air pollution analysis in Delhi, India," Atmospheric Environment, vol. 67, pp. 101–111, 2013.

43.   S. Dey, L. D. Girolamo, A. V. Donkelaar, S. N. Tripathi, T. Gupta, and M. Mohan, "Variability of outdoor fine particulate ($PM_{2.5}$) concentration in the Indian Subcontinent: a remote sensing approach,"Remote Sensing of Environ, vol. 127, pp. 153–161, 2012.

44.   K. He, F. Yang, Y. Ma et al., "The characteristics of $PM_{2.5}$ in Beijing, China," Atmospheric Environment, vol. 35, no. 29, pp. 4959–4970, 2001.

45.   J. G. Watson, "Visibility: science and regulation," Journal of the Air and Waste Management Association, vol. 52, no. 6, pp. 628–713, 2002.

46.   S. F. Rajši , M. D. Tasi , V. T. Novakovi , and M. N. Tomaševi , "First assessment of the PM_{10} and $PM_{2.5}$ particulate level in the ambient air of Belgrade City," Environmental Science and Pollution Research, vol. 11, no. 3, pp. 158–164, 2004.

47.   EPCA report number 9 (November 2004) Report on the increase in the number of three-wheelers in Delhi, In response to the Hon'ble Supreme Court Order Dated October 8, 2004, In response to the I.A. 217 of 2003.

48. N. Singh, Role of atmospheric ions on condensation and cloud formation processes [Ph.D. thesis], University of Roorkee, Roorkee, India, 1985.

49. R. R. Ranjan, H. P. Joshi, and K. N. Iyer, "Spectral variation of total column aerosol optical depth over Rajkot: a tropical semi-arid Indian station," Aerosol and Air Quality Research, vol. 7, no. 1, pp. 33–45, 2007.

50. P. C. S. Devara and P. E. Raj, "A lidar study of atmospheric aerosols during two contrasting monsoon seasons," Atmosfera, vol. 11, no. 4, pp. 199–204, 1998.

51. J. Wallace and P. Kanaroglou, "The effect of temperature inversions on ground-level nitrogen dioxide (NO_2) and fine particulate matter ($PM_{2.5}$) using temperature profiles from the Atmospheric Infrared Sounder (AIRS)," Science of the Total Environment, vol. 407, no. 18, pp. 5085–5095, 2009.

52. D. M. Chate and P. C. S. Devara, "Growth properties of submicron aerosols during cold season in India," Aerosol and Air Quality Research, vol. 5, no. 2, pp. 127–140, 2005.

53. S. Cheng and K. Lam, "An analysis of winds affecting air pollution concentrations in Hong Kong," Atmospheric Environment, vol. 32, no. 14-15, pp. 2559–2567, 1998.

54. A. Chaloulakou, P. Kassomenos, N. Spyrellis, P. Demokritou, and P. Koutrakis, "Measurements of PM_{10} and $PM_{2.5}$ particle concentrations in Athens, Greece," Atmospheric Environment, vol. 37, no. 5, pp. 649–660, 2003.

55. S. Ruellan and H. Cachier, "Characterisation of fresh particulate vehicular exhausts near a Paris high flow road," Atmospheric Environment, vol. 35, no. 2, pp. 453–468, 2001.

Chapter 5

Satellite and Ground Measurements for Studying the Urban Heat Island Effect in Cyprus

Diofantos G. Hadjimitsis[1], Adrianos Retalis[2], Silas Michaelides[3], Filippos Tymvios[3], Dimitrios Paronis[2], Kyriacos Themistocleous[1], and Athos Agapiou[1]

[1]Cyprus University of Technology, Faculty of Engineering and Technology, Department of Civil Engineering and Geomatics, Remote Sensing and Geo-Environment Laboratory, Cyprus

[2]National Observatory of Athens, Greece

[3]Cyprus Meteorological Service, Cyprus

INTRODUCTION

An urban heat island (UHI) is a phenomenon whereby an urban area experiences elevated air temperatures due to anthropogenic modification of the environment and is usually more evident at night. During heat waves the local effect of an UHI is superimposed on the regional temperature field and as a result heat stress is enhanced. Both the intensity and the spatial structure of the observed thermal contrast of the UHI depend on various parameters, such as the structure of the urban tissue, the population density and its associated heat release, the land use patterns, the vegetation cover, the surface topography and relief etc. In general terms, the UHI is becoming more intense as city sizes increase. Traditional measurements of the near-surface UHI are based on measurements of the air temperature using urban and rural weather stations or air temperature transects. Thermal satellite sensors, which primarily measure the radiance at the top of the atmosphere in the thermal infrared, retrieve the so called land surface temperature (LST) which is the temperature measured at the Earth's surface and is regarded as its skin temperature. Given that LST is different from the surface air temperature, a distinction is made in remote sensing studies between surface urban heat island (SUHI) and atmospheric heat island (e.g., Nichol, 1996).

Several studies published in the literature have focused on the use of remotely sensed data for studying the urban heat island effect (Dousset & Gourmelon, 2003; Kato & Yamaguchi, 2005; Lo & Quattrochi, 2003; Streutker, 2002; Tran et al., 2006; Xiao et al., 2007; Yuanbo et al., 2007). Other relevant studies are focusing on the validation of satellite LST retrievals with ground measurements or on the inter-comparison of LST products from different sensors (Mostovoy et al., 2005; Nichol et al., 2009; Retalis et al., 2010). The availability of a multitude of data archives (e.g., from MODIS, ASTER and Landsat TM/ETM+ sensors) with long time-series has recently raised the scientific interest in the relevant field. As a result, several studies have been published on the study of the UHI effect for various cities of the world (Hung et al. 2006; Imhoff et al., 2010; Peng et al., 2012).

This Chapter discusses the urban heat island effect in Cyprus based on both multi-temporal satellite and meteorological data. The necessary information of the study area is provided in Section 2. The description and selection of the heat waves and the analysis of the synoptic conditions favouring the development of heat waves are discussed in Section 3. The development of a Neural Network for the correlation of satellite derived land surface temperature (LST) with ground based air surface temperature is examined in Section 4. The analysis of satellite derived LST for studying the temporal evolution of LST and the deviation of LST (anomaly) from the mean values during a heat wave event are presented in Section 5, while Section 6 refers to the calculation of the mean monthly magnitude of urban heat island (UHI) for the period 2002-2008 and for selected heat wave events.

STUDY AREA

The island of Cyprus is located in the eastern part of the Mediterranean Basin. The island is situated between latitudes circles 34° and 36° N, and meridians 32° and 35°E. Cyprus has a typical eastern Mediterranean climate: the combined temperature–rainfall regime is characterized by cool-to-mild wet winters and warm-to-hot dry summers (Michaelides et al., 2009). The climatological annual precipitation of Cyprus is around 500mm. The highest precipitation is recorded in the mountainous areas with 1100mm, while in the coastal areas precipitation is limited to 300-350mm.

From a morphological point of view, the island can be divided into five main morphological regions: (a) The mountainous complex of Troodos located at the center of Cyprus; (b) the mountain range of the Pentadaktylos at the northern part; (c) the central plain of Mesaoria located between of these two mountainous ranges; (d) the hilly areas around the mountainous complex of Troodos; and (e) the coastal plains (see Fig. 1). The coastline of Cyprus is characterized by numerous capes and bays. The narrow coastal plains in the north are covered with olive trees and carob trees, while a short distance from the coast, the northern mountain range (Pentadaktylos) is found, which is a limestone formation and peaks to a height of 1024 meters. At the south and the east of the island there are two salt – lakes.

The Troodos mountain range with a peak at 1951 m covers most of the south-western part and the center of the island. This area is covered almost by forests, mainly pine and other forest trees such as cypresses, oaks and cedars. It is estimated that forests cover about 19% of the total area of the island.

Cyprus is divided into six districts: Kyrenia, Famagusta, Larnaca, Limassol, Paphos and Nicosia (Fig. 2). During the last decade (2000-2010), there has been recorded a dramatic urban expansion (see Fig. 3). As it was found from previous studies (Hadjimitsis et al., 2011), there has been an increase of urban areas of more than 100% compared to late 1980's and a decrease of 20% of rural areas. These results were derived from an analysis of multi-temporal satellite image classification.

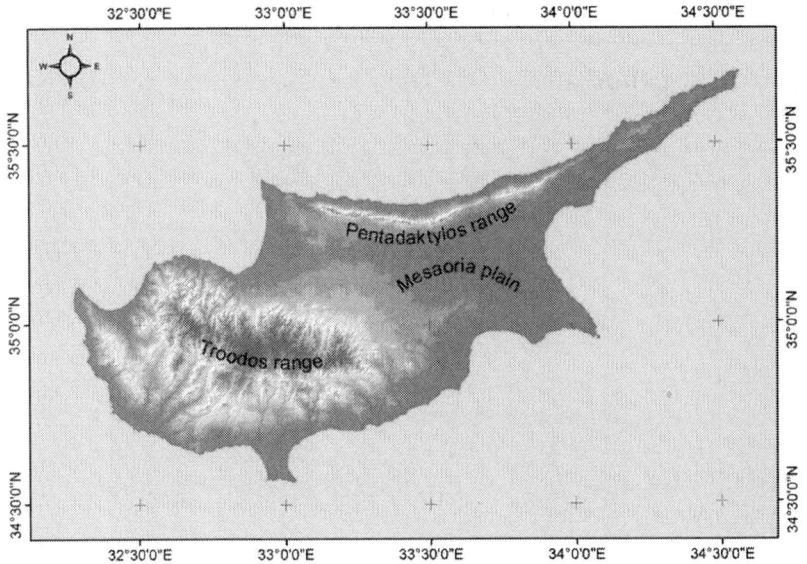

Figure 1: Main morphological regions of Cyprus.

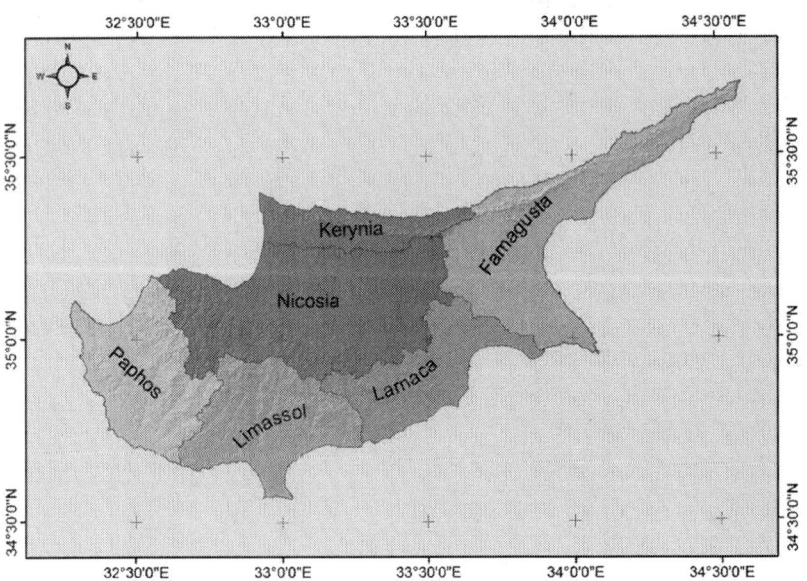

Figure 2: Districts of Cyprus.

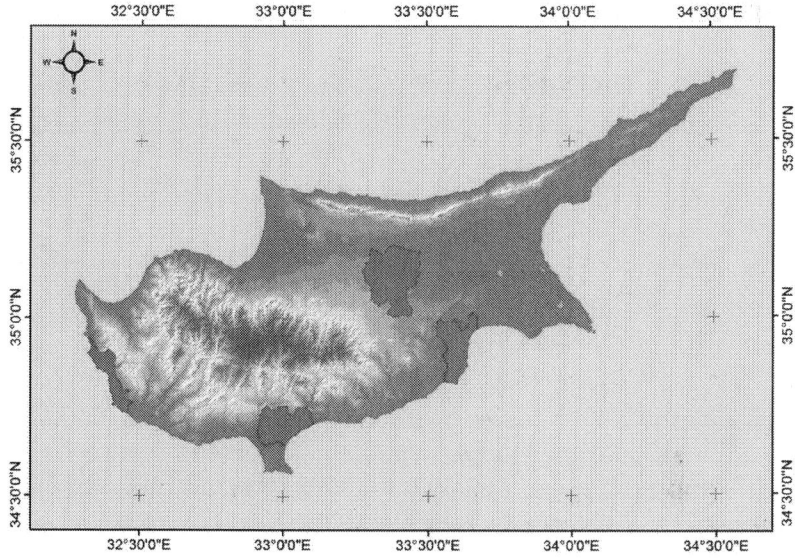

Figure 3: Urban areas of Cyprus shown in red.

HEAT WAVE EVENTS AND SYNOPTIC PATTERNS

A strong relationship exists between large scale circulation patterns and regional surface variables such as surface pressure, dynamical rainfall, wind and temperature (Tymvios et al., 2007, 2008, 2010a;Xoplaki et al., 2003). As a consequence, synoptic upper air charts at certain levels comprise a valuable tool for the operational weather forecaster to qualitatively predict occurrences of weather phenomena observed on the ground (e.g., heavy rainfall; see Tymvios et al., 2010a). The height pattern at 500 hPa is often used for this purpose. In order to take advantage of these semi-empirical methods and to simplify the statistical processing, stochastic downscaling methods are often applied to the actual weather patterns in order to generate clusters of synoptic cases with similar characteristics. Weather type classifications are simple, discrete characterizations of the atmospheric conditions and they are commonly used in atmospheric sciences. For a review of various classifications, including their applications, refer to Key &

Crane (1986), El-Kadi & Smithoson (1992), Hewitson & Crane (1996) andCannon & Whitfield (2002).

Heat waves have a distinct impact on society through increased mortality, change in the energy consumption profile and the diversification of social behavior. The severity of the heat events may include the local climatological characteristics, the community design and the individual tolerance to heat. Both the frequency of appearance and the intensity of heat waves are increasing in the Mediterranean area (Founda & Giannakopoulos, 2009).

The eastern Mediterranean climate is characterized by the succession of a single rainy season (November to mid-March) and a single longer dry season (mid-March to October). This generalization is modified by the influence of maritime factors, yielding cooler summers and warmer winters in most of the coastal and low-lying areas. Visibility is generally very good. However, during spring and early summer, the atmosphere is quite hazy, with dust transferred by the prevailing south-easterly to southwesterly winds from the Saharan and Arabian deserts, usually associated with the development of desert depressions (Michaelides et al., 1999). The influence of synoptic types on the urban heat island has been investigated by Mihalakakou et al. (2002) who have also adopted a neural network approach.

The definition for a heat wave recommended by the World Meteorological Organization is "when the daily maximum temperature of more than five consecutive days exceeds the maximum temperature normal by 5°C". Nevertheless, in most countries, the definition of extreme heat events is based on the potential for hot weather conditions to result in an unacceptable level of adverse health effects, including increased mortality. Also, a threshold in maximum temperature is in practical use in many countries.

These periods of abnormally and uncomfortably hot and (usually) humid weather are very common in the eastern Mediterranean during summer and early autumn. Expert examination of the synoptic patterns on upper air charts may reveal the potential for a heat wave event. In this respect, the research presented here attempts to identify height patterns favorable for heat events by using a neural network classification method, namely, Kohonen's Self Organizing Maps (see Kohonen, 1990).

Data

As an indication of a possible heat event, the maximum temperature of Nicosia station in Cyprus was chosen. This station is located within the urban area of the city of Nicosia (35°17'N, 33°35'E, 170m, see Fig. 4) and equipped with traditional instrumentation was operational from 1957 until 2001, when it was upgraded to an automatic station. The database used in this study comprises the maximum and minimum temperature records from this station. The maximum monthly temperature measurements are presented in Fig. 5. Also, for the classification of synoptic patterns, the ERA40 reanalysis for the period of 1958 to 2000 (covering roughly the ERA40 time window) were utilized.

The temperatures database was checked for consistency and homogeneity against measurements from nearby stations while the maximum temperatures were also checked for normal distribution fitting.

Methodology

The maximum daily temperature at Nicosia station was checked against the climatological monthly average value of the period 1961-1990. If the difference was 5°C or more, then the period was characterized as "possible heat event". If the subsequent days were also positive against this temperature test for more than three days, then the period was considered as heat event. The heat events were checked against the weather classification patterns in order to identify a connection among particular patterns and heat events. The same procedure was adopted for a difference of 3°C, since events with a 5°C difference are rare even during summer. Special care was taken when checking the last and the first day of the month whereby daily maximum temperature values were subtracted from the average climatological value of the two subsequent months.

Figure 4: Location of ground stations used.

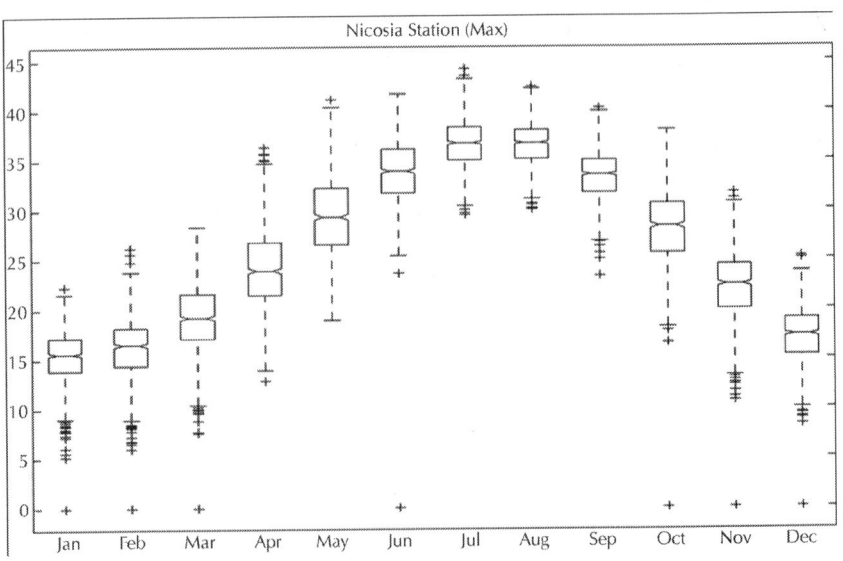

Figure 5: Box and Whiskers plot of the maximum temperatures in Nicosia (1958 - 2000).

Details of the Self Organizing Maps methodology used for the classification have been presented inMichaelides et al. (2010). The 36-Cluster classification adopted also in the present study has been recently demonstrated by Tymvios et al. (2010b).

Results

The distribution of the heat events in consecutive days for 3°C and 5°C difference is illustrated in Fig. 6. It is clearly evidenced that more than 75% of the events last for 3 to 5 days. Most of the identified heat events occur in the transition periods (i.e., Spring and Autumn). This finding is also supported by the findings in Fig. 5, where the larger variation (the area between 25th and 75th percentile) of the average of the maximum temperatures is given for the same periods. With the exception of the periods 12 to 21 July 1978 (10 days) and 2 to 14 July 2000 (13 days), all incidents lasting more than 10 days for this station occurred in October, November, March, April and May.

Clusters 5 and 34 share most of the heat event occurrences. They are both transition period clusters with similar characteristics, exhibiting a distinctive upper level ridge over the eastern Mediterranean and a deep low to the west of this ridge; Cluster 5 belongs to the cold period and Cluster 34 to the warm period. An example of a Cluster 5 member is illustrated in Fig. 7.

When these clusters appear during early Spring and late Autumn, the heat events last from 8 to 15 days, while when they appear just before or after Summer (May and September) they last around 5 days. Summertime appearances of heat events are equally shared between Clusters 12, 19, 24 and 36, all characterized by warm and dry conditions (Michaelides et al., 2010).

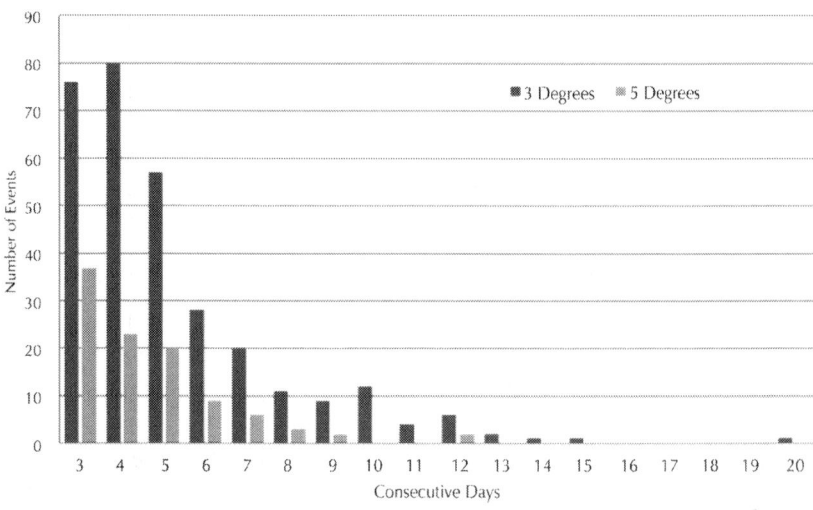

Figure 6: Distribution of the heat events in consecutive days for 3°C and 5°C difference.

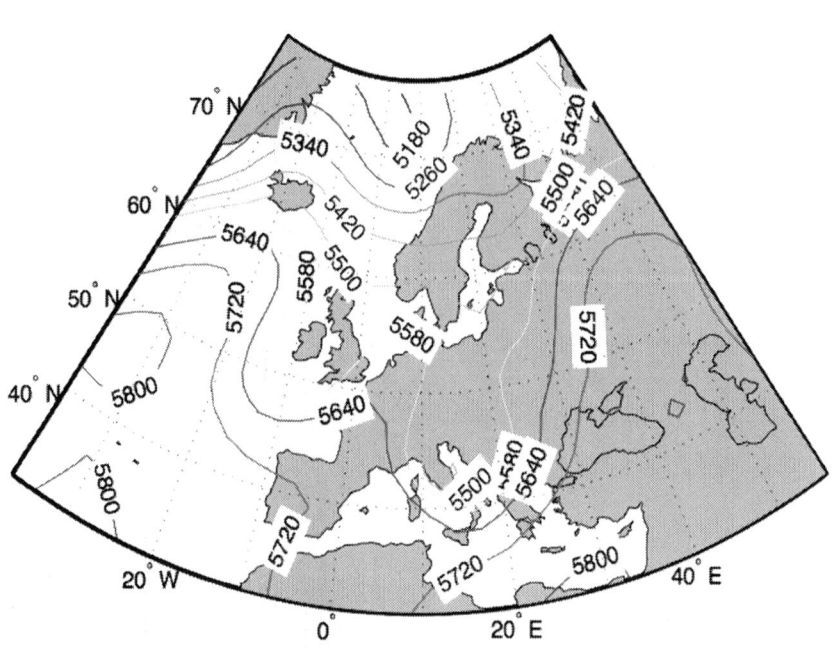

Figure 7: The height pattern at 500hPa (Cluster 5) from 24 November 1962.

Although it appears that some Clusters are associated with heat events over Cyprus, the connection between heat events and atmospheric circulation at 500hPa did not give definite results that any of these patterns dominate heat event occurrences (as it was possible to demonstrate in previous studies on rainfall and extreme rainfall events). There might be two reasons for this inadequacy. The first is that the window that was chosen for the classification does not include the synoptic patterns that influence the area sufficiently; the second reason is that, although upper air patterns at 500hPa contribute significantly to the evolution of certain surface features (such as dynamical or extreme rainfall), such an association is not so clear for the temperature field. In the search for associations of the temperature fields with synoptic patterns in the Mediterranean, it is important to consider also the lower parts of the atmosphere.

Future research concerning the connection of the weather classification patterns will be focused into a new, much larger window that will include Northern Africa and the Middle East and a combination of classification of patterns at lower levels of the atmosphere (e.g., 850hPa, 700hPa).

SATELLITE ESTIMATES OF TEMPERATURE VERSUS GROUND MEASUREMENTS

In this Section, a methodology is presented in which the temperature estimates from the MODIS sensor onboard the Terra Satellite is contrasted with ground measurements. The methodology consists of a neural network approach in which measurements on the ground are used as input to the neural network, whereas, the temperature estimate from the satellite is considered as the network's output.

The neural network methodology adopted has successfully been implemented before in tackling several climatological problems in Cyprus: the prediction of maximum daily total solar irradiance (Kalogirou et al., 2002), the prediction of the daily average solar radiation (Tymvios et al., 2002, 2005a), the modeling of photosynthetic radiation (Tymvios et al., 2005b) and others.

Data

For the needs of this research, data from MODIS onboard the Terra satellite have been used. More specifically, the level-3 product MOD11A1 (version 5) for the period 2000-2009 was exploited, at a resolution of about 1km by 1km (0.927km). Using the available Land Surface Temperature (LST) fields derived from MODIS, a time series was established corresponding to the position of ground stations. Wan & Dozier (1996) have developed the Generalized Split Window (GSW) algorithm for the retrieval of LST, using the thermal (infrared) channels of MODIS and under different atmospheric conditions (see also, Wan, 1999, 2008). This algorithm retrieves LST on the basis of emissivities in bands 31 and 32 of MODIS. The accuracy in estimating LST was found to be better than 1K, whereas in most cases it was better than 0.5K (Hulley & Hook, 2009; Coll et al., 2009).

The data base for the surface measurements used in this research consist of the hourly recorded temperature at each of the automatic meteorological stations of the network operated by the Cyprus Meteorological Service (see Fig. 4), in the period 2000-2007. Based on these data, the maximum temperature recorded in the time period 1100 – 1300 UTC (local standard time=UTC+2 hours) was considered as the day's maximum and was subsequently used in the study.

The training of the neural network implemented requires that there are no missing data in the time series used, because the data are used in groups and are therefore inter-dependent. Therefore, the estimated LST (by the neural network implemented) is based on the data of a whole day and missing values result in the rejection of that day. Following quality control based on the above constraint, the number of automatic stations that were subsequently used was reduced to twelve, as shown in Table 1.

Methodology

Artificial Neural networks (ANN) are small autonomous computational units (algorithms) with inter-connections which, to a large extent, resemble the functioning of natural computational units, namely, the neurons of the human brain. ANN can be trained and learn through repeated examples so that they can reach conclusions and results without human intervention. Since their invention, ANN covered a wide spectrum of research and disciplines and their application has been phenomenal. A few of the numerous examples of ANN applications are mentioned here: medical systems' automation for the recognition of malignant tumors, control of military equipment and aircraft, estimation of environmental variables, quality verification in production factories, forecasting of financial indices, weather diagnosis and forecasting etc.

For the implementation of the ANN methodology in the present research, the Multi-Layer Perceptron (MLP) was adopted (see Haykin, 1998). The input to this network is the surface temperature recorded at the ground stations and the output is the temperature estimated by the satellite (LST). Fig. 8 displays the MLP implemented.

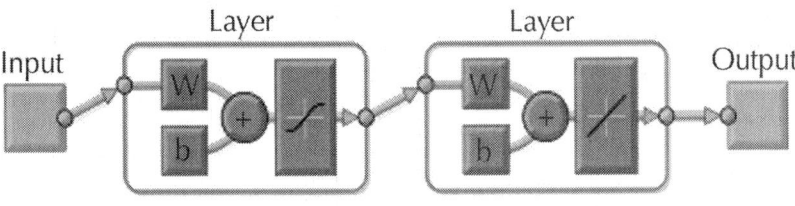

Figure 8: The Multi-Layer Perceptron (MLP) network implemented for the prediction of LST.

The data from the twelve ground stations and the respective MODIS estimations of LST were used as follows: 60% were used for Training of the network, 20% for Validation and the remaining 20% as an independent set for Testing.

Results

Table 1 displays the errors in the estimation of LST with the neural network, by using the independent set of data. In this table, the maximum, minimum and average errors along with the standard deviation are shown for each of the ground stations. Overall, the performance of the neural network is considered as very satisfactory. However, there are cases where the error is unacceptable and this requires further investigation.

The relation between input data (ground temperature) and satellite estimated temperature LST (target) is shown in Fig. 9. The results are shown for all the data but also separately for the Training, Validation and Testing data sets. For the Training set, the correlation coefficient is R=0.96991, for the Validation set R=0.89692 and for the Testing set R=0.9145, whereas, for all the data R=0.94747. Based on these findings, the performance of the network in predicting LST is considered as satisfactory.

Table 1: Errors of LST estimation for the independent set of data for each ground station

Ground station	Latitude (N)	Longitude (E)	Altitude Above sea level (m)	Maximum (°C)	Minimum (°C)	Average (°C)	Standard deviation (°C)
Astromeritis	35°03′	32°26′	175	6.81	-7.31	0.22	2.35
Athalassa	35°04′	33°58′	162	7.14	-7.58	0.63	2.77
Athienou	35°03′	33°32′	185	6.45	-5.74	0.23	2.27
Dasaki	35°03′	33°47′	50	5.74	-4.52	0.60	2.17
Kannaviou	34°55′	32°35′	419	7.03	-6.23	0.48	2.41
Kathikas	34°55′	32°26′	650	8.25	-5.30	0.63	2.53
Kato Pyrgos	35°11′	32°41′	5	7.19	-10.12	0.62	2.99
Malia	34°49′	32°47′	645	7.59	-7.75	-0.11	2.56
Mennogeia	34°51′	33°26′	140	7.36	-7.78	0.28	3.07
Paphos	34°47′	32°26′	82	8.34	-7.45	0.37	2.83
Paralimni	35°04′	33°58′	65	7.51	-8.42	0.68	2.61
Polis	35°03′	32°26′	20	6.87	-4.83	0.40	2.40

In this research, an attempt has been made to relate ground measurements of temperature with the temperature as it is estimated from MODIS and develop a neural network methodology that can be used in the estimation of ground temperatures by using the satellite imagery. Although the methodology performs sufficiently, overall, it seems that further refinement is needed in order to improve the approach. The adoption of a single network for all the time series of data seems to limit the application of the methodology. For example, the present single neural network developed for each station does not take into account the large seasonal variations in the parameter concerned. It could be more effective if several neural networks are developed based on seasonally grouped data.

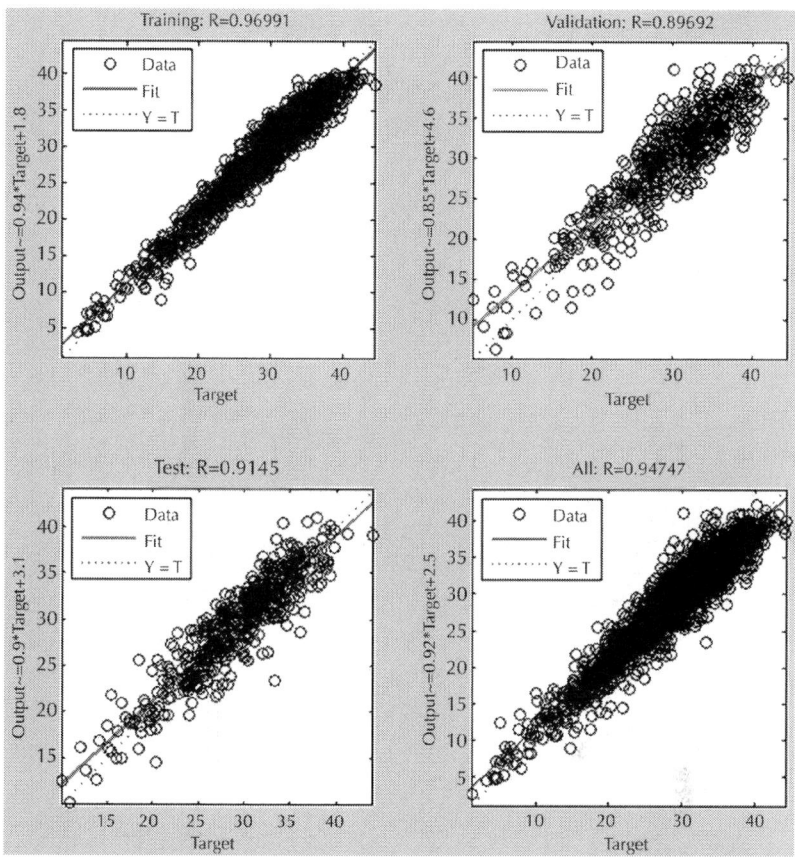

Figure 9: Neural network performance for the Training, Validation, Test and All data sets.

LAND SURFACE TEMPERATURE ANALYSIS

The MODIS sensor, onboard Terra and Aqua polar satellites, provides one day and one night image under clear sky conditions. MODIS is particularly suitable for the land surface temperature (LST) product due to its global coverage, radiometric resolution and dynamic ranges for a variety of land cover types and high calibration accuracy in multiple thermal bands.

MODIS LST product is based on the generalized split-window (GSW) algorithm (Wan & Dozier, 1996) using as input the MODIS thermal bands 31 and 32. The parameters in the MODIS GSW depend on the satellite zenith view angles, column water vapor and also on the low atmosphere boundary temperature. The band emissivities rely on the classification-based method (Snyder et al., 1998) according to land cover types in the pixel (Monteiro et al., 2007). Temperatures are extracted in Kelvin; accuracy of 1 Kelvin is yielded for materials with known emissivities (Wan, 1999), while a number of studies have also tested the accuracy of the MODIS LST product with favorable results (Wan, 2002; Wan et al., 2004; Coll et al., 2005; Wan, 2008).

The MODIS Aqua product MYD11A1 (V5) and MODIS Terra product MOD11A1 (V5) – Land Surface Temperature and Emissivity Daily L3 Global 1 km Grid SIN were used. Terra and Aqua overpass times for the study area are considered at approximately 1030 and 1330 UTC for day passes, and at approximately 2230 and 0130 UTC for night passes, respectively.

The use of MODIS LST data for examining the temporal evolution and the retrieved temperature anomaly maps for a heat wave event occurred on 24 June 2007 is presented. Moreover, MODIS LST data are used for calculating the urban heat island (UHI) at four urban areas of Cyprus during the extreme heat wave of August 2010.

Modis LST Temporal Evolution and Temperature Anomaly Maps

MODIS LST data were initially used for generating mean monthly climatology LST maps for June in the period of 2003-2008. The mean and maximum Aqua day and night LST values for June are presented in Fig. 10 for the period 2003-2008 for two urban areas (Nicosia, Larnaca) and one rural area (Ag. Marina). The curves show that the mean night LST values for the two urban areas are similar, while for the area of Ag. Marina, the temperature levels are 3-4 °C lower. For all sites, a minimum was observed for year 2005. The situation is different though regarding day LST values. The coastal site of Larnaca exhibited the lowest values among the three areas, while Nicosia and Ag. Marina exhibited similar patterns and temperature levels. The overall trend over time for the three areas showed a positive trend.

The intense heat wave event of 24 June 2007 was next examined in order to study the LST behavior during such events since satellite derived LST is controlled by land cover and topographic effect factors. In Fig.11, temperature anomaly maps, in terms of temperature deviation from the long-term monthly mean values (calculated for the period 2003-2008), are presented for the heat wave event under consideration and for both night and day Aqua passes.

The spatial patterns observed in the temperature anomaly maps are complex. It can be observed that day LST anomaly is more intense (up to 15°C) than the night anomaly. Minimum anomaly is located in the area of the mountain range Troodos (central-eastern part), while the southern part of Cyprus presents higher anomaly values than the northern part. The different values of LST increase are attributed to the difference in the emitted radiance from each land type and/or the urban heat island effect.

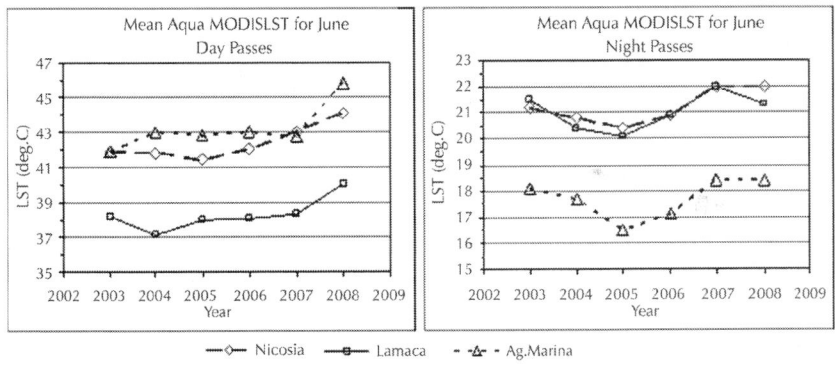

Figure 10: Yearly evolution of the mean Aqua MODIS LST for June (2003-2008) as retrieved from Aqua satellite for three different areas in Cyprus.

The amplitude of LST anomaly variation between day and night was examined with the land cover types based on the CORINE 2000 land cover map (Fig. 12). It was found that the mean anomaly amplitude was 2.89-4.05°C for artificial surfaces, 2.87-6.01°C for agricultural areas and 2.81-4.63°C for forest and semi natural areas. However, variations were noticed even in the same category. For example, for artificial surfaces the higher amplitude was noticed for airports and the lower for dump sites. For agricultural areas, the higher amplitude was noticed for pastures and the lower for annual crops associated with permanent crops. For forest and semi natural areas, the higher amplitude was noticed for beaches, dunes and sands and the lower for mixed forest.

A close inspection on the Aqua LST image (Fig. 12) acquired on 24 June 2007 (day pass) depicted that the highest LST values are noticed in areas that are recognized as vulnerable to desertification (Fig. 13). In Cyprus, there are two climatic zones that are considered as sensitive to desertification: the semi-arid, which extends over the larger part of the island and the arid sub-humid, which covers the slopes of the Troodos range and the windward side and higher parts of the Kyrenia range (IACO, 2007).

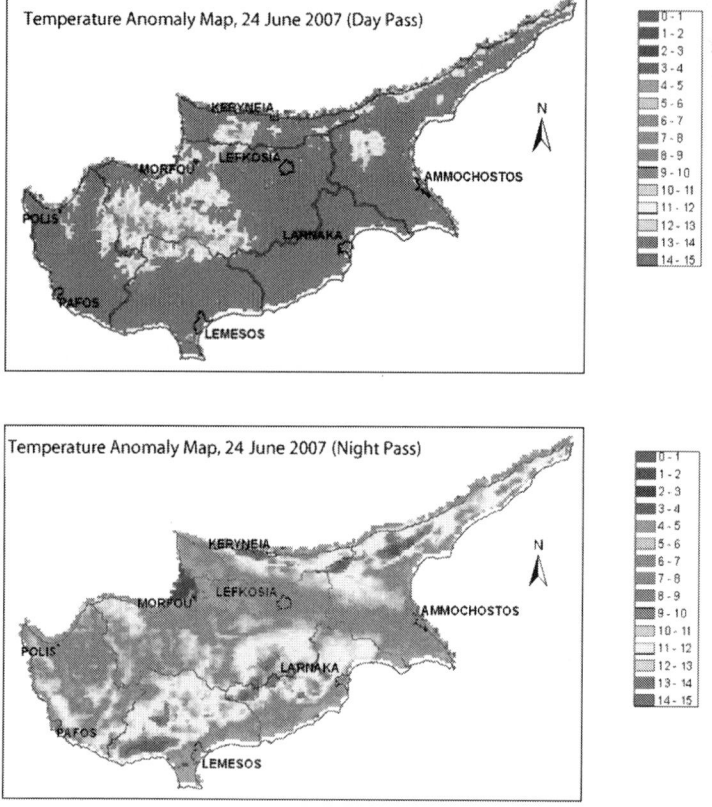

Figure 11: Land Surface Temperature anomaly map derived from both day (top) and night (bottom) Aqua MODIS passes for the selected heat wave event.

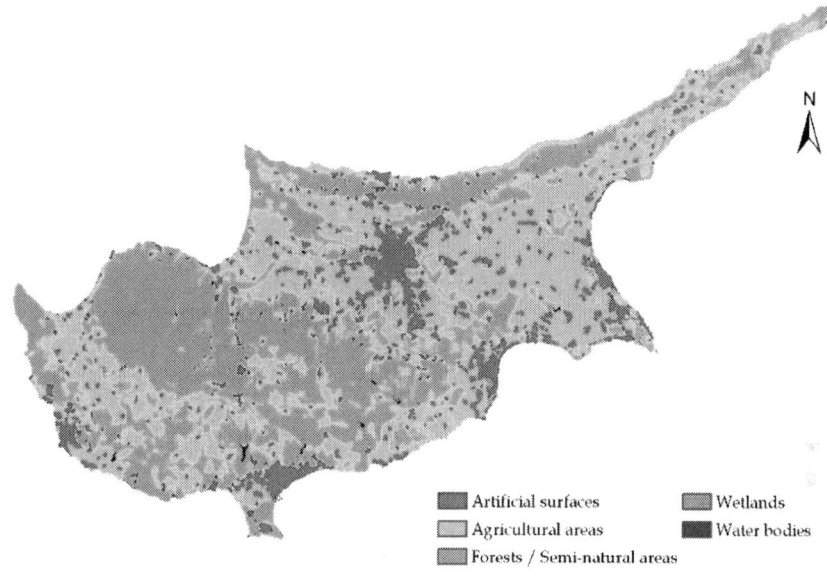

Figure 12: Simplified CORINE 2000 Land Cover map of Cyprus.

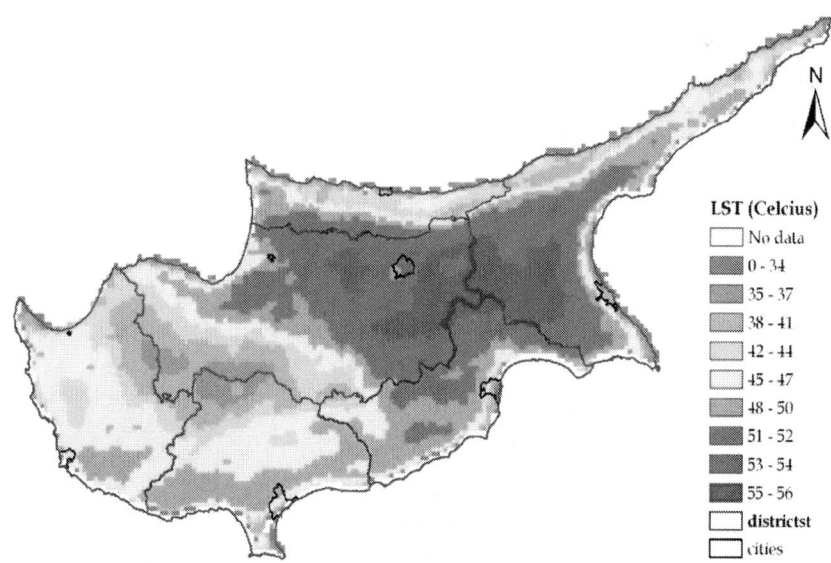

Figure 13: Land Surface Temperature map derived from day Aqua pass for the selected heat wave event.

URBAN HEAT ISLAND ANALYSIS

The variation of the UHI magnitude was examined for the four urban areas of Cyprus based on MODIS Aqua images acquired at night-time (at approximately 0130 local time). The selection of the MODIS Aqua data was based on the criterion that the night-time acquired images allow a more precise LST calculation since there is no incoming solar radiation to change the surface radiation balance, while night-time MODIS LST accuracy has been found to be better than day time (Rigo et al., 2006).

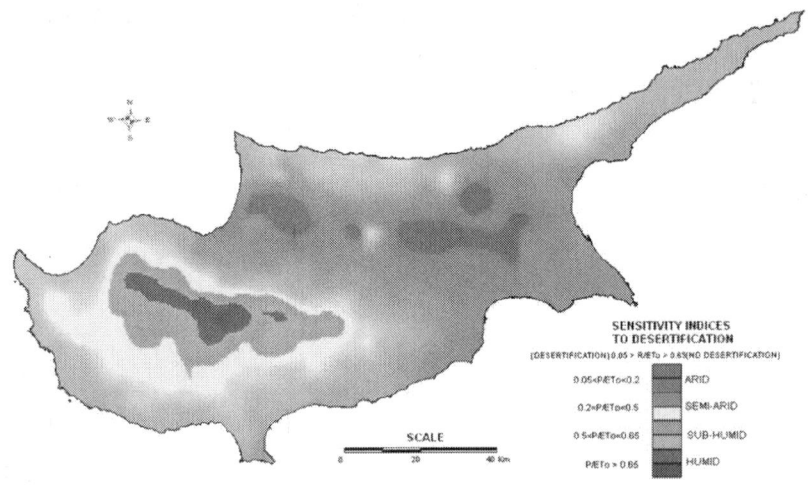

Figure 14: Areas sensitive to desertification, according to the United Nations Convention to Combat Desertification (IACO, 2007).

The magnitude of the UHI was estimated for each of the four test sites both for the mean monthly period 2002-2008 (Fig. 14) and for selective days of high temperature records of August 2010 (Fig 15). The UHI magnitude was calculated by subtracting the LST value from a rural area (as identified from the position of a pre-selected rural meteorological station) from the respective LST values falling within the urban boundary area of each district on a pixel-by-pixel basis (Tomlinson et al., 2010).

Fig.14 presents the mean monthly maximum UHI intensity for the period 2002-2008 for the four urban areas of Cyprus. As noticed,

Nicosia, which is located in the centre of Cyprus, is most vulnerable to UHI during the warm period, when the intensity is recorded above four degrees. On the contrary, the other urban areas (Larnaca, Limassol and Paphos), which are close to the coastline, are lesser affected by UHI during the warm period, with intensities recorded around 1.5 to 3.5°C. These areas also demonstrated high UHI intensities during the cold period.

Next, the spatio-temporal variation of the UHI intensity for each of the urban areas was examined for the period 23 July to 28 August 2010, when high air surface temperatures were recorded (Fig. 14). The temporal variation of the maximum UHI intensity was estimated from the available nocturnal Aqua MODIS images for that period. The results revealed that, for most of the cases, the UHI magnitude curves follow a similar trend. Two major peaks were observed, on 31 July and 25 August 2010.

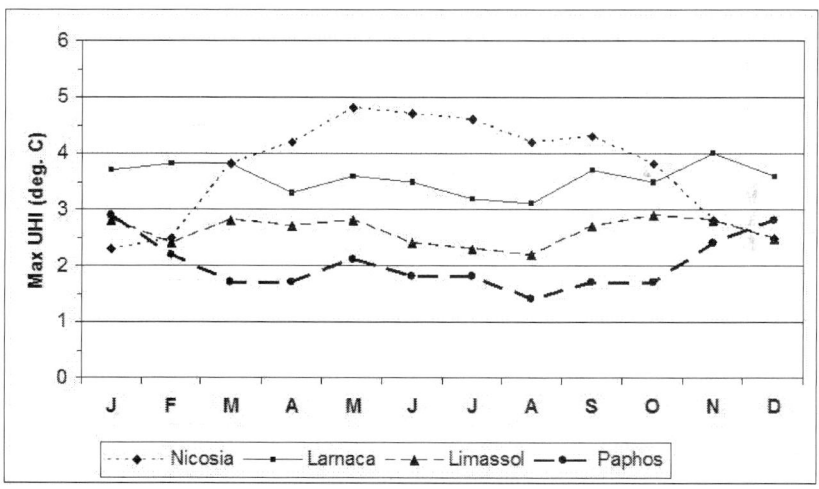

Figure 15: Mean monthly maximum UHI magnitude estimated from MODIS Aqua nocturnal images for the period 2002-2008 for Nicosia, Larnaca, Limassol and Paphos.

The spatial variation of the UHI magnitude (Fig. 16) was examined for two dates (31 July and 28 August 2010) and was compared to the mean UHI magnitude as calculated for August for the years 2002-2008. The results derived suggest that, in almost all cases, the spatial patterns of the UHI magnitude observed for each urban area are quite similar to each other with a few variations in the magnitude of intensity due to the severity of the heat wave event. The highest intensities were noticed within the areas of the urban fabric.

The maximum intensities of UHI for each urban area were (a) 31 July 2010: 5.2°C (Nicosia), 3.5°C (Larnaca), 1.9°C (Limassol), and 5.0°C (Paphos) and (b) 25 August 2010: 6.9°C (Nicosia), 3.9°C (Larnaca), 3.1°C (Limassol), and 4.2°C (Paphos). Thus, the deviation form the mean monthly UHI intensities calculated for July and August, correspondingly, were of about 0.6°C and 2.7°C for Nicosia, 0.3°C and 0.8°C for Larnaca, -0.4°C and 1.9°C for Limassol and 3.2°C and 2.8°C for Paphos.

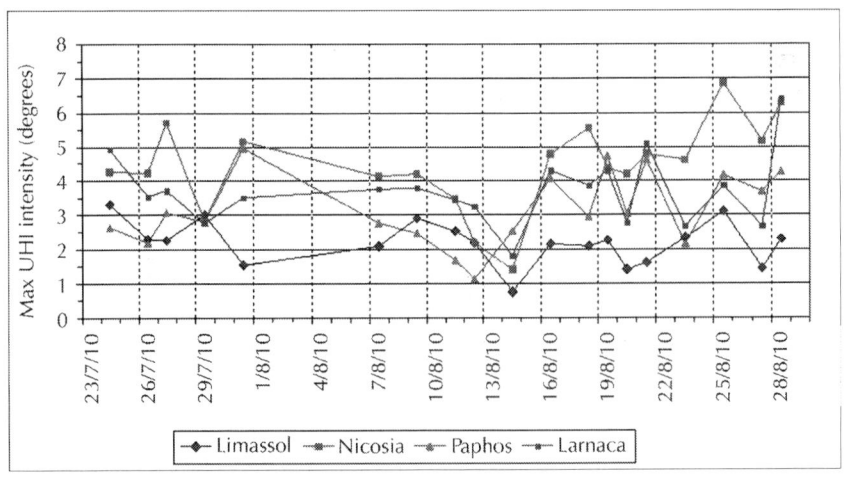

Figure 16: Temporal variation of maximum UHI intensity for the four urban areas of Cyprus, as derived from the analysis of Aqua nocturnal data for the period 23 July to 28 August 2010.

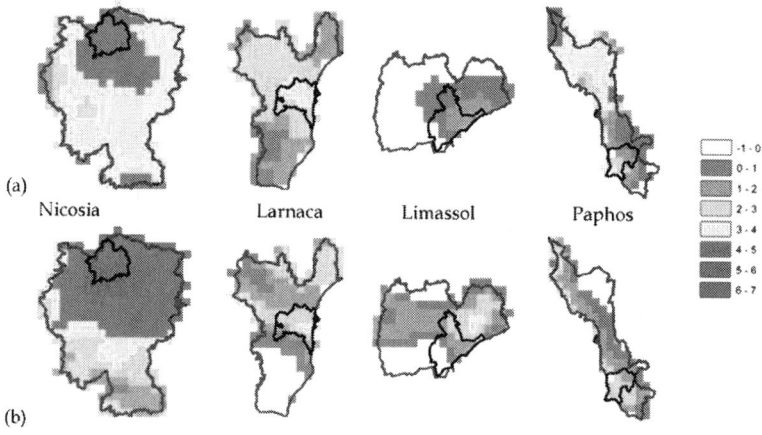

Figure 17: UHI estimated from MODIS Aqua nocturnal images for (a) 31 July and (b) 28 August 2010, for the four urban areas of Cyprus, separately.

CONCLUDING REMARKS

This Chapter commented on the use of Earth observation data along with ground meteorological data for the study of the UHI phenomenon in Cyprus. The synoptic conditions favoring the development of heat wave events were discussed. Neural Network analysis was used for classifying synoptic patterns and relate them with heat events. The majority of the heat events have occurred during the transition periods (Spring and Autumn). However, despite the fact that some clusters can be associated with such phenomena the connection between these events and atmospheric circulation at 500hPa did not give clear results.

Furthermore, an attempt was made in order to correlate ground temperature measurements and MODIS LST data. The results have shown that the methodology can perform sufficiently; however, further refinement is needed in order to improve this approach.

Aqua MODIS retrievals of land surface temperature data were used for studying selective heat wave events. The analysis of LST data depicted the regions that are more prone to such events. The spatial variations of the UHI magnitude was also examined for the major cities of Cyprus, during both mean monthly conditions and for selected events, identifying areas that are most vulnerable.

ACKNOWLEDGEMENTS

The results presented in this Chapter form part of the research project "Study of the Phenomenon of Urban Heat Island in Cyprus", funded by the Cyprus Research Promotion Foundation of Cyprus, under contract No. AEIFORIA/ASTI/0308(BE)/01. Cyprus Meteorological Service is kindly acknowledged for the provision of the meteorological data. MODIS LST data were distributed by the Land Processes Distributed Active Archive Center (LP DAAC), located at the U.S. Geological Survey (USGS) Earth Resources Observation and Science (EROS) Center (lpdaac.usgs.gov).

REFERENCES

1. Cannon, J.A. & Whitfield, P.H. (2002). Synoptic map-pattern classification using recursive partitioning and principal component analysis, *Monthly Weather Review*, Vol.130, pp. 1187–1206

2. Coll C.; Caselles V.; Galve J.; Valor E.; Niclos R. ; Sanchez J. & Rivas R. (2005). Ground measurements for the validation of land surface temperatures derived from AATSR and MODIS data, *Remote Sensing of Environment,* Vol.97, pp. 288–300, DOI:10.1016/j.rse.2005.05.007

3. Coll, C.; Wan, Z. & Galve J.M. (2009). Temperature-based and radiance-based validations of the V5 MODIS land surface temperature product,*Journal of Geophysical Research*, 114, D20102, doi:10.1029/2009JD012038

4. Dousset, B. & Gourmelon, F. (2003). Satellite multi-sensor data analysis of urban surface temperatures and landcover, *ISPRS,* 58, pp. 43–54

5. El-Kadi, A.K.A. & Smithoson, P.A. (1992). Atmospheric classifications and synoptic climatology, *Progress in Physical Geography*, 16, pp.432–455

6. Founda, D. & Giannakopoulos, C. (2009). The exceptionally hot summer of 2007 in Athens, Greece: A typical summer in the future climate, *Global and Planetary Change*, 67, pp.227-236, doi:10.1016/j.gloplacha.2009.03.013

7.   Hadjimitsis D. & Agapiou A. (2011). A remote sensing evaluation of urban expansion and its impact on Urban Heat Island phenomenon: the case study of Limassol, Cyprus, *Proceedings of the 5th International Conference Earth From Space - the Most Effective Solutions*, 29/11 - 01/12/2011, Moscow, Russia, p.301-303, ISBN:978-5-9518-0490-7

8.   Haykin, S. (1998). *Neural Networks: A Comprehensive Foundation*, 2nd edition, Macmillan College Publishing

9.   Hewitson, B.C. & Crane, R.G. (1996). Climate downscaling: Techniques and application. *Climate Research*, 7, pp.85–95

10. Hulley, G.C. & Hook, S.J. (2009). Intercomparison of versions 4, 4.1 and 5 of the MODIS Land Surface Temperature and Emissivity products and validation with laboratory measurements of sand samples from the Namib desert, Namibia, *Remote Sensing of Environment*, 113, pp. 1313-1318

11. Hung, T.; Uchihama, D.; Ochi, S. & Yasuoka, Y. (2006). Assessment with satellite data of the urban heat island effects in Asian mega cities, *International Journal of Applied Earth Observation and Geoinformation*, 8, pp. 34−48

12. IACO (2007). *Consultation Services for the Production of a National Action Plan to Combat Desertification in Cyprus, I*, IACO Environmental and Water Consultants Ltd, Cyprus.

13. Imhoff, M. L., Zhang, P., Wolfe, R. E., Bounoua, L. (2010). Remote sensing of the urban heat island effect across biomes in the continental USA, *Remote Sensing of Environnent*, 114 (3), pp. 504−513

14. Kalogirou, S.; Michaelides, S. & Tymvios, F. (2002). Prediction of maximum solar radiation using artificial neural networks. *World Renewable Energy Congress VII (WREC 2002)*, Cologne, 29 June – 5 July, 2002, A.A. Sayigh (Ed.), Elsevier Science Ltd (on CD-ROM)

15. Kato, A. & Yamaguchi, Y. (2005). Analysis of urban heat-island effect using ASTER and ETM+ Data: separation of anthropogenic heat discharge and natural heat radiation from sensible heat flux, *Remote Sensing of Environment*, 99, pp. 44–54

16. Key, J. & Crane, R. J. (1986). A Comparison of Synoptic classification schemes based on "objective" procedures, *Journal of Climatology*, 6, pp. 375–388

17.   Kohonen, T. (1990). The Self-Organizing Map, *Proceedings of the IEEE*, 78, pp.1464–1480

18.   Lo, C.P. & Quattrochi, D.A., (2003). Land-use and land-cover change, urban heat island phenomenon, and health implications: a remote sensing approach, *Photogrammetric Engineering & Remote Sensing*, Vol.69, pp. 1053–1063

19.   Monteiro, I.; Trigo, I.F.; Kebasch, E. & Olesen F. (2007). Validation of land surface temperature retrieved from Meteosat Second Generation Satellites, *Proceedings of the Joint 2007 EUMETSAT Meteorological Satellite Conference and the 15th Satellite Meteorology & Oceanography Conference of the American Meteorological Society*, Amsterdam, The Netherlands, 24-28 September 2007, Available from: http://www.eumetsat.int/Home/Main/Publications/index.htm

20.   Mostovoy, G.V.; King, R.; Reddy, K.R. & Kakani, V.G. (2005). Using MODIS LST data for high-resolution estimates of daily air temperature over Mississippi, *Proceedings of the International Workshop on the Analysis of Multi-Temporal Remote Sensing Images*, 16-18 May 2005, pp. 76-80

21.   Michaelides, S.C. ; Evripidou, P. & Kallos, G. (1999). Monitoring and predicting Saharan desert dust transport in the eastern Mediterranean,*Weather*, Vol.54, pp.359–365

22.   Michaelides, S. ; Tymvios , F. & Michaelidou, T. (2009). Spatial and temporal characteristics of the annual rainfall frequency distribution in Cyprus, *Atmospheric Research,* Vol.94, pp. 606–615

23.   Michaelides, S.; Tymvios, F.S. & Charalambous, D. (2010). Investigation of trends in synoptic patterns over Europe with artificial neural networks,*Advances in Geosciences*, Vol.23, pp.107-112

24.   Mihalakakou, G.; Flocas, H.A. ; Santamouris, M. & Helmis, C.G. (2002). Application of Neural Networks to the Simulation of the Heat Island over Athens, Greece, Using Synoptic Types as a Predictor, *Journal of Applied Meteorology*, Vol.41, pp. 519-527

25.   Nichol, J.E. (1996). High resolution surface temperature patterns related to urban morphology in a tropical city: a satellite-based study, *Journal of Applied Meteorology,* Vol.35, pp. 135–146

26. Nichol, J. E.; Fung, W. Y.; Lam, K. & Wong, M. S. (2009). Urban heat island diagnosis using ASTER satellite images and 'in situ' air temperature,*Atmospheric Research*, Vol.94, pp. 276–284

27. Peng, S.; Piao, S.; Ciais, P.; Friedlingstein, P.; Ottle, C.; Bréon, F-M.; Nan, H.; Zhou, L. & Myneni, R. B. (2012). Surface Urban Heat Island across 419 global big cities, *Environmental Science & Technology*, Vol.46, pp. 696-703, dx.doi.org/10.1021/es2030438

28. Retalis, A.; Paronis, D. ; Lagouvardos, K. & Kotroni, V. (2010). The heat wave of June 2007 in Athens, Greece—Part 1: Study of satellite derived land surface temperature, *Atmospheric Research*, Vol.98, pp. 458–467

29. Rigo, G.; Parlow, E. & Oesch D. (2006). Validation of satellite observed thermal emission with in-situ measurements over an urban surface, *Remote Sensing of Environment,* Vol.104, pp. 201–210, doi: 10.1016/j.rse.2006.04.018

30. Snyder, W. C.; Wan, Z.; Zhang, Y. & Feng, Y.-Z. (1998). Classification-based emissivity for land surface temperature measurement from space.*International Journal of Remote Sensing,* Vol.19, pp. 2753–2774

31. Streutker, D.R. (2002). A remote sensing study of the urban heat island of Houston, Texas, *International Journal of Remote Sensing,* Vol.23, pp. 2595–2608

32. Tomlinson, C.; Chapman, L.; Thornes J. & Baker C. (2010) Derivation of Birmingham's summer surface urban heat island from MODIS satellite images, *International Journal of Climatology,* doi:10.1002/joc.2261

33. Tran, H.; Uchihama, D.; Ochi, S. & Yasuoka, Y. (2006). Assessment with satellite data of the urban heat island effects in Asian mega cities,*International Journal of Applied Earth Observation and. Geoinformation,* Vol.8, pp. 34–48

34. Tymvios, F.; Jacovides, K.P. & Michaelides, S.C. (2002). Calculation of global solar radiation on a horizontal surface by using Artificial Neural Networks, *Proceedings of the 6th Panhellenic Conference on Meteorology, Climatology and Atmospheric Physics,* Ioannina, Greece, 25-28 September, 2002 (In Greek) pp. 468-475

35. Tymvios, F.S.; Jacovides, C.P. ; Michaelides, S.C. & Skouteli, C. (2005a). A comparative study of ngstrom›s and artificial neural

networks> methodologies in estimating global solar radiation, *Solar Energy*, Vol.78, pp.752-762

36.   Tymvios, F.; Jacovides, C. ; Michaelides, S. ; Schizas, C. & Scouteli, C. (2005b). Modelling the photosynthetically active radiation (PAR) for Nicosia in Cyprus, *Proceedings of the 7th Panhellenic (International) Conference of Meteorology, Climatology and Atmospheric Physics*, 28-30 Sept., 2004, Nicosia, Cyprus, pp. 972-980

37.   Tymvios, F.S. ; Constantinides, P. ; Retalis, A. ; Michaelides, S. ; Paronis, D. ; Evripidou, P. & Kleanthous S. (2007). The AERAS project– database implementation and Neural Network classification tests, *Proceedings of the 6th International Conference on Urban Air Quality*, Limassol, Cyprus, 27-29 March 2007

38.   Tymvios F.S.; Savvidou K.; Michaelides S.C. & Nicolaides K.A. (2008). Atmospheric circulation patterns associated with heavy precipitation over Cyprus, *Geophysical Research Abstracts*, Vol.10, EGU2008-A-04720

39.   Tymvios, F.; Savvidou, K. & Michaelides, S. (2010a). Association of geopotential height patterns with heavy rainfall events in Cyprus, *Advances in Geosciences*, Vol.23, pp.73-78

40.   Tymvios F.; Charalambous D ; Michaelides S. ; Retalis A. ; Paronis D. & Skouteli C. (2010b). Temperature distribution in Cyprus with the use of satellite images and artificial neural networks, *Proceeding of the 10th International Conference on Meteorology, Climatology and Atmospheric Physics*, 25-28 May 2010, Patra, Greece, 227-234

41.   Wan Z. & Dozier J. (1996). A generalized split-window algorithm for retrieving land-surface temperature from space, *IEEE Transactions on Geoscience and Remote Sensing*, Vol.34, pp. 892–905, doi:10.1109/36.508406

42.   Wan Z. (1999). *MODIS Land-Surface Temperature Algorithm Theoretical Basis Document (LST ATBD)*, University of California, Santa Barbara, USA, Institute for Computational Earth System Science, Available from: http://modis.gsfc.nasa.gov/data/atbd/atbd mod11.pdf

43.   Wan Z. (2002). Validation of the land-surface temperature products retrieved from Terra Moderate Resolution Imaging Spectroradiometer data, *Remote Sensing of Environment*, Vol.83, pp. 163–180, doi:10.1016/S0034-4257(02)00093-7

44.   Wan Z. ; Zhang Y. ; Zhang Q. & Li Z. (2004). Quality assessment and validation of the MODIS global land surface temperature, *International Journal of Remote Sensing,* Vol.25, pp. 261–274, doi:10.1080/0143116031000116417

45.   Wan, Z. (2008). New refinements and validation of the MODIS land-surface temperature/emissivity products, *Remote Sensing of Environment,* Vol.112, pp.59-74

46.   Xiao, R.; Ouyang, Z.; Heng, H.; Li, W.; Schienke, E. & Wang, X. (2007). Spatial pattern of impervious surfaces and their impacts on land surface temperature in Beijing, China. *Journal of Environmental Science* , Vol.19, pp. 250–256

47.   Xoplaki, E.; Gonzalez-Rouco, J.F.; Luterbacher, J. & Wanner, H. (2003). Mediterranean summer air temperature variability and its connection to the large-scale atmospheric circulation and SST, *Climate Dynamics,* Vol.20, pp.723-739

48.   Yuanbo Liu, Y.; Yamaguchi, Y. & Ke, C. (2007). Reducing the discrepancy between ASTER and MODIS Land Surface Temperature products,*Sensors,* Vol.7, pp. 3043–3057

Chapter 6

Initialization of Tropical Cyclones in Numerical Prediction Systems

Eric A. Hendricks[1] and Melinda S. Peng[1]

[1]Marine Meteorology Division, Naval Research Laboratory, Monterey, CA, USA

INTRODUCTION

Tropical cyclones (here after TCs) are intense atmospheric vortices that form over warm ocean waters. Strong TCs (called hurricanes in the North Atlantic basin or typhoons in the western north Pacific basin) can cause significant loss of lives and property when making landfall due to destructive winds, torrential rainfall, and powerful storm surges. In order to warn people of hazards from incoming TCs, forecasters must make predictions of the future position and intensity of the TC.

In order to make these forecasts, a forecaster uses a wide suite of tools ranging from his or her subjective assessment of the situation based on experience, the climatology and persistence characteristics of the storm, and most importantly, *models*, which make a prediction of the future state of the atmosphere given the current state. In this chapter, the focus is on dynamical models. A dynamical model is based on the governing laws of the system, which for the atmosphere are the conservation of momentum, mass, and energy. Since the system of partial differential equations that govern the atmosphere is highly nonlinear, a numerical approximation must be made in order to obtain a solution to these equations. Short term (less than 7 days) numerical weather prediction is largely an initial value problem. Therefore it is critical to accurately specify the initial condition. The accuracy of the initial condition depends on the forecast model itself, the quality and density of observations, and how to distribute the information from the observations to the model grid points (data assimilation). Since most TCs exist in the open oceans, most observations come from satellites, and often intensity and structure characteristics are inferred from the remotely sensed data [10]. Therefore a key problem that remains for TC initialization is the lack of observations, especially in the inner-core (less than 150 km from the TC center).

TCs are predicted using both global and regional numerical prediction models. Global models simulate the atmospheric state variables on the sphere, while regional model simulate the variables in a specific region, and thus have lateral boundaries. Due to smaller domains of interest, regional models can generally be run at much higher horizontal resolution than global models, and thus they are more useful for predicting tropical cyclone intensity and structure. As an example of how well TC track and intensity has historically been predicted, Fig. 1 shows the average track and intensity errors from official forecasts from the National Hurricane Center from 1990-2009. While there has been a steady improvement in the ability to predict track (left panel), there has been little to no improvement in this time period in the prediction of TC intensity (right panel). Currently there is a large effort to improve intensity forecasts: the National Oceanic and Atmospheric Administration (NOAA) Hurricane Forecast Improvement Project (HFIP).

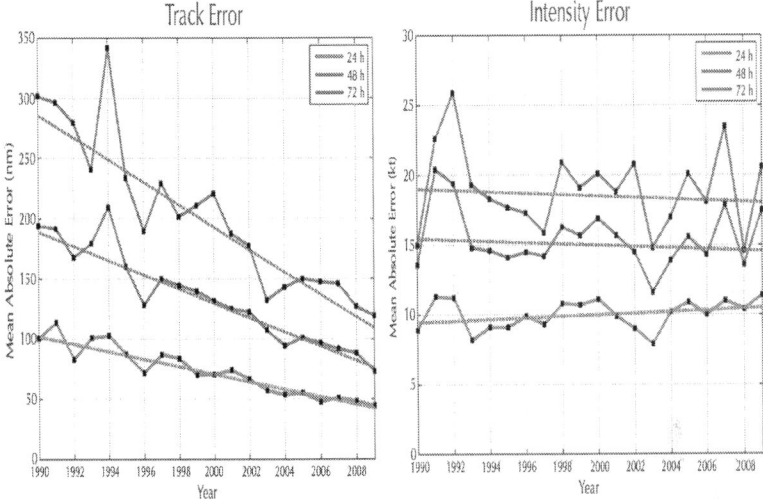

Figure 1: Average mean absolute errors for official TC track (left panel) and intensity (right panel) predictions at various lead times in the North Atlantic basin from 1990-2009. Data is courtesy of the National Hurricane Center in Miami, FL, and plot is courtesy of Jon Moskaitis, Naval Research Laboratory, and Monterey, CA.

Errors in the future prediction of TC track, intensity and structure in numerical prediction systems arise from imperfect initial conditions, the numerical discretization and approximation to the continuous equations, model physical parameterizations (radiation, cumulus, microphysics, boundary layer, and mixing), and limits of predictability. While improvements in numerical models should be directed at all of these aspects, in this chapter we are focused on the initial condition. The purpose of TC initialization is to give the numerical prediction system the best estimate of the observed TC structure and intensity while ensuring both vortex dynamic and thermodynamic balances. In this chapter, a review of different types of TC initialization methods for numerical prediction systems is presented. An overview of the general TC structure and challenges of initialization is given in the next section. In section 3, the direct vortex insertion schemes are discussed. In section 4, TC initialization methods using variational and ensemble data assimilation systems are discussed. In section 5, initialization schemes that are designed for improved initial balance are discussed. A summary is provided in section 6.

OVERVIEW OF THE TC STRUCTURE

Tropical cyclones come in a wide variety of different structures and intensities. Intensity is a measure of the strength of the TC, and is usually given in terms of a maximum sustained surface wind or the minimum central pressure. Structure is a measure of various axisymmetric and asymmetric features of the TC in three dimensions. Structure encompasses the outer wind structure (such as the radius of 34 kt wind), inner core structure (such as the radius of maximum winds, eyewall width and eye width), as well as various asymmetric features (inner and outer spiral rain bands, asymmetries in the eyewall, asymmetric deep convection, and asymmetries due to storm motion and vertical wind shear). Additionally, structure would encompass vertical variations in the TC (such as the location of the warm core and how fast the tangential winds decay with height). While there are some observations (particularly for horizontal aspects of the structure from remote satellite imagery), there are never enough observations to know the complete three-dimensional flow and mass field in the TC.

In this section we outline some important structural aspects of the TC, including the basic axisymmetric and asymmetric structures that should be incorporated into the numerical model initial condition. An atmospheric state variable, which may be temperature or velocity, may be interpolated to a polar coordinate system about the TC center and decomposed as

$\psi\ (r,\ \square,\ p,\ t) = \overline{\psi}\,(r,\ p,\ t) + \psi'(r,\ \square,\ p,\ t)$, where $\overline{\psi}\,(r,\ p,\ t)$ is the axisymmetric component of the variable (where the overbar denotes as azimuthal mean), and $\psi'(r,\ \square,\ p,\ t)$ is the asymmetric component of the variable. Here r is the radius from the vortex center, $\square$ is the azimuthal angle, p is the pressure height, and t is the time. Often TCs are observed to be mostly axisymmetric (but with lower azimuthal wavenumber asymmetries due to storm motion and vertical shear); however in certain instances, and in certain regions of the TC, there can be large amplitude asymmetric components.

Axisymmetric Structure

Fig. 2 shows the basic axisymmetric structure of a TC from a real case, Hurricane Bill (2009), obtained from the initial condition of

(COAMPS®) numerical prediciton system [1] - shown. In the Fig. 2a, the azimuthal mean tangential velocity is shown, in Fig. 2b the radial velocity is shown, and in Fig. 2c the perturbation temperature is shown. There are three important regimes in Fig. 2: (i) the boundary layer, (ii) the quasi-balance layer, and (iii) the outflow layer. The boundary layer is the region of strong radial inflow near the surface in Fig. 2b. Above the boundary layer, the winds are mostly tangential in the quasi-balance layer, and then at upper levels (Fig. 2b) the outflow layer with strong divergence and radial outflow is evident. In Fig. 2a, it can be seen that the strongest tangential winds are near the surface and decay with height, and in Fig. 2c a mid to upper level warm core is evident. While this is just one case, it illustrates the basic axisymmetric structure of a TC. While the vertical velocity is not shown in this figure, there exists upward motion in the eyewall region, and this combined with the low to mid-level radial inflow and upper level outflow constitute the hurricane's secondary (or transverse) circulation. Changes in the secondary circulation are largely responsible for TC intensity change.

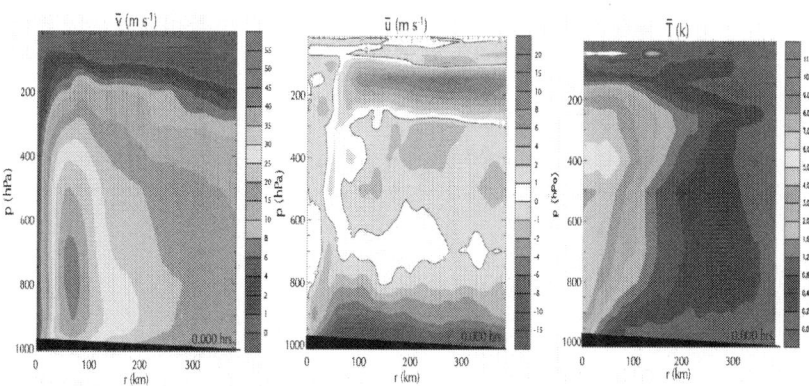

Figure 2: Azimuthal mean structure of the initial condition of Hurricane Bill (2009) in the Naval Research Laboratory's Coupled Ocean/Atmosphere Mesoscale Prediction System COAMPS®. Panels: a) tangential velocity (m s^{-1}), b) radial velocity (m s^{-1}), and c) perturbation temperature (K). Reproduced from [18] © Copyright 2011 AMS (http://www.ametsoc.org/pubs/crnotice. html).

Using the quasi-balance approximation, where the vorticity is much larger than the divergence, the f-plane radial momentum equation can be approximated by

$$\frac{\partial \Phi}{\partial r} = \frac{v^2}{r} + fv,$$

(1)

Where $\Phi = gz$ is the geopotential, v is the tangential velocity, f is the Coriolis parameter, and r is the radius from the TC center. Outside of deep convective regions, the hydrostatic approximation (in pressure coordinates) is also largely valid,

$$\frac{\partial \Phi}{\partial p} = -\frac{RT}{p},$$

(2)

Where p is the pressure, R is the gas constant, and T is the air temperature. Taking $\partial/\partial p$ (1) and $\partial/\partial r$ (2) while eliminating the mixed derivative term, the vortex thermal wind relation is obtained

$$\frac{\partial v}{\partial p}\left(\frac{2v}{r} + f\right) = -\frac{R}{p}\frac{\partial T}{\partial r}.$$

(3)

This equation states that a vortex in which v decreases with decreasing p must have warm core, i.e., T must decrease with increasing radius. This is evident in Fig. 2b, where the warm core begins at upper levels, where v is rapidly decreasing.

In the outflow and boundary layers, there exists significant divergent and convergence, respectively, such that the quasi-balance approximation is no longer valid. Therefore an appropriate initialization scheme for TCs should not only capture the primary axisymmetric tangential (azimuthal) circulation, but also the secondary circulation, including the boundary and outflow layers. Additionally, there must be a thermodynamic balance between the boundary layer inflow, rising air in deep and shallow convection, and upper level outflow.

Asymmetric Structure

In order to illustrate some asymmetric features in TCs, Fig. 3 shows two hurricanes: Hurricanes Dolly (2008) and Alex (2010). Hurricane Dolly was very asymmetric in the inner-core region. Note the azimuthal wavenumber-4 pattern in the eyewall radar reflectivity. Hurricane Alex (2010) was also very asymmetric, and had a large spiral rainband emanating from the core, and no visible eye. The point illustrated here is that TCs come in a wide variety of shapes and sizes, and often have prominent asymmetric features. While there is some structure dependence on intensity (i.e., stronger TCs in general are more axisymmetric than weaker TCs), at any initial time a given TC may have very different structure, and the goal of the initialization system is to capture its true state. Remote satellite measurements generally give a decent estimate of the horizontal structure. In fact, microwave data has allowed the ability to see through visible and infrared cloud shields, giving improved estimates of the deep convection and precipitation. However, typically there is much less data about the vertical structure. For example, the boundary layer structure or convective and stratiform heating profiles of Alex's rainband would not generally be known. Due to the lack of observations in TCs, in TC initialization systems, aspects of the structure are often specified using estimated information from satellite images.

Figure 3: Radar and visible satellite imagery depicting asymmetric features in TCs. Hurricane Dolly (2008) (left panel) had asymmetries in the eyewall and rain bands. Hurricane Alex (2010) (right panel) had a large azimuthal wave-number-1 spiral rain band propagating outward from the vortex center. The left panel is courtesy of the NOAA National Weather Service and the right panel is courtesy of the NOAA/NESDIS in Fort Collins, CO.

DIRECT INSERTION SCHEMES

As discussed in the previous section, TCs are poorly observed, particularly in the inner-core region. The North Atlantic basin is the only basin that routinely has aircraft reconnaissance missions into storms when they are close to the U.S. southeast coastal regions. The aircraft reconnaissance missions can provide important inner-core structural data using airborne Doppler radar and dropwindsondes, as well as direct or remote measurements of surface wind speed and minimum central pressure. Due to the lack of observations of the inner-core structure of TCs, vortex bogussing has been used to improve the representation of the TC in numerical prediction systems. Generally speaking, vortex bogussing is the creation of a TC-like vortex that can be inserted into the initial fields of numerical models [28]. The direct insertion methods take a bogus vortex and insert it directly into the numerical model initial conditions. The bogus vortex can be generated in different ways, which are described below. The main strength of these methods is that the vortex is usually self-consistent. However, some weaknesses exist. First, there can be imbalances that may exist when blending the inserted vortex with the environments in the model analysis. Secondly, for weak TCs and TCs experiencing vertical shear, it is not desirable to insert a vertically stacked vortex into the initial conditions (which is often the case with bogus vortices). Additionally previous studies have shown strong sensitivity to the vertical structure of the bogus vortex, which is often not well observed [46].

After a bogus vortex is created, there needs to be a method to properly insert this vortex into the initial fields of the forecast model. The first guess fields (or the previous model forecast which is valid at the analysis time), usually will already contain a TC-like vortex from the previous forecast. However this vortex may have an incorrect position, intensity, and structure, and therefore it should be removed from model fields. Vortex removal and insertion methods require a number of steps. The common method, discussed by [26] is as follows. First, the total field (e.g., surface pressure) is decomposed into a basic field and disturbance field using filtering. Next, the vortex with specified length scale is removed from the disturbance field. Then, the environmental field is constructed by adding the non-hurricane disturbance with the basic field. Finally, the specified vortex can then simply be added to the environmental field. Schemes of this nature are widely used in operational tropical cyclone prediction models in order to improve the TC representation from the global analysis [27, 34, and 50].

Static Vortex Insertion

Since TCs are observed to largely be in gradient and hydrostatic balance above the boundary layer [49], one method is to insert a balanced vortex. Routine warning messages are generated by TC warning centers that include estimates of the maximum sustained surface wind, central pressure, and size characteristics (such as the radii of 34 kt winds). Using a function fit to the observed radial wind profile (e.g., a modified Rankine vortex or more sophisticated methods [19, 20]) along with a vertical decay assumption, one can obtain an axisymmetric tangential wind field in the radius-height plane. Following this, the mass field (temperature and pressure) may be obtained by solving the nonlinear balance equation in conjunction with the hydrostatic equation. Then this balanced vortex may be directly inserted into the model initial conditions, as a representation of the actual observed TC vortex. While this method is relatively straightforward, there are a few potential problems: (i) TC vortices are not balanced in the boundary and outflow layers, where strong divergence exists, and (ii) in convectively active regions of the vortex the hydrostatic balance assumption is not valid. It is possible to relax the strict balance assumptions above by building in the boundary layer and outflow structure diagnostically. The addition of boundary and outflow layers should reduce the amount of initial adjustment after insertion.

Insertion of a Dynamically Initialized Vortex

Instead of specifying a vortex (usually analytically) to represent a TC, another method is to spin-up a TC-like vortex in a numerical model in an environment with no mean flow, and then insert this vortex into the model initial conditions. This method is called a TC dynamic initialization method because the TC vortex is developed from numerical simulation of a nonlinear atmospheric prediction model with full physics that requires prior model integration. The benefits of such a procedure are that the numerical model will generate a more realistic structure for the boundary layer and the outflow layer, and the moisture variables can also be included. The TC dynamic initialization is usually accomplished through Newtonian relaxation. A Newtonian relaxation term is added to the right hand side of a desired prognostic variable (e.g., the tangential velocity or surface pressure) in order to anchor the vortex to the desired structure and/or intensity. The Geophysical Fluid Dynamics Laboratory hurricane prediction model uses an axisymmetric version of its primitive equation to perform the dynamic initialization to a prescribed structure [3, 26, and 27]. Recent work has also shown encouraging results with the TC dynamic initialization method using an independent three-dimensional primitive equation model in conjunction with a three-dimensional variational (3DVAR) data assimilation scheme [18, 61]. In Fig. 4, a flow diagram is shown depicting a TC dynamic initialization method applied after three-dimensional variational (3DVAR) data assimilation, where TCs are spun up using Newtonian relaxation to the observed surface pressure. This procedure showed a positive improvement in TC intensity prediction, as average errors in maximum sustained surface wind and minimum central pressure was reduced at all forecast lead times.

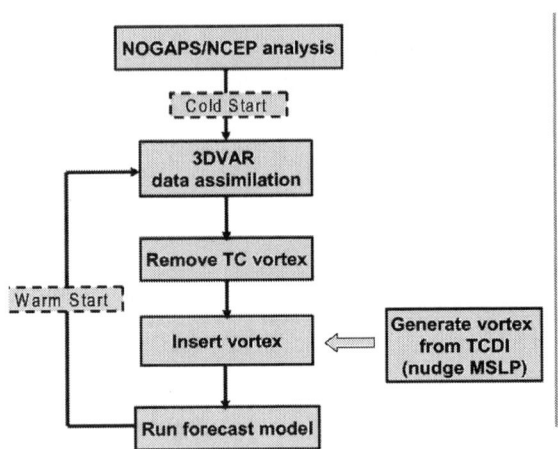

Figure 4: Application of a TC dynamic initialization scheme to a 3DVAR system, reproduced from [18]. A TC is nudged to observed central mean sea level pressure (MSLP) in a nonlinear full-physics model, and then inserted into the forecast model initial conditions after 3DVAR © Copyright 2011 AMS (http://www.ametsoc.org/pubs/crnotice.html).

DATA ASSIMILATION SYSTEMS FOR TC INITIALIZATION

The purpose of data assimilation is to produce initial states (analyses) for numerical prediction that maximizes the use of information contained in observations and prior model forecasts to produce the best possible predictions of future states. Most data assimilation methods use observations (e.g., in-situ and remote measurements) to correct short-term model forecasts (the first guess), and therefore the accuracy of the resulting analysis is not just a function of the data assimilation methodology, but the fidelity of the forecast model itself. This analysis is then used as the initial condition for the forecast model. In this section, we discuss the data assimilation strategies that incorporate observational data into the model for proper representation of TCs at the initial time.

In the variational method, a cost function is minimized to produce an analysis that takes into account both the model and observation (including instrument and representativeness) errors. 3DVAR systems (or three-dimensional variational methods) solve this cost function in the three spatial dimensions, while 4DVAR (four-dimensional) systems add the temporal component in a set window. Generally speaking, most atmospheric observations are more applicable to the synoptic scale flow pattern, and often there are few (if any) observations of the inner-core of TCs or other mesoscale or small scale phenomena, aside from infrequent field campaigns. Yet even if these observations exist, it is not trivial to assimilate them while ensuring the proper vortex dynamic and thermodynamic balances.

3DVAR Systems

The replacement of optimal interpolation (OI) data assimilation scheme by the variational (VAR) method significantly improved the forecast skill of numerical weather prediction systems. The motivation originated from the difficulties associated with the assimilation of satellite data such as TOVS (TIROS-N Operational Vertical Sounders) radiances. It was shown by [31] that the statistical estimation problem could be cast in a variational form (3DVAR) which is a different way of solving the problem than the OI scheme which solves directly. The first implementation of 3DVAR was done at the National Centers for environmental Prediction (NCEP) [36] and later on at the European Center for Medium Range Weather Forecasting (ECMWF) [4]. Other centers like the Canadian Meteorological Centre [13], the Met Office [30], and Naval Research Laboratory [6] also implemented a 3DVAR scheme operationally.

The common method for TC vortex initialization in 3DVAR systems is through the use of adding synthetic observations [15, 17, 29, 55, and 65]. Synthetic observations are observations that are created from the estimates of the TC structure and intensity that come from tropical cyclone warning centers (such as the National Hurricane Center in Miami, FL, and the Joint Typhoon Warning Center in Pearl Harbor, HI), and give the best estimate of the storm position, intensity and structure. The synthetic observations are used to enhance the TC representation in the numerical model initial conditions, which generally cannot be adequately captured using the conventional observations. The synthetic observations themselves may be created by sampling a function that matches the observed vortex, and these observations are treated as radiosonde data with assigned proper position information and are included with all other observations and blended with the model first guess using the 3DVAR system. Generally speaking, the observation error is set very low with the TC synthetic observations in the assimilation process, so that the analysis process will largely retain these characteristics of the synthetic observations near the TC. A number of TC synthetic observations are shown for Typhoon Morakot (2009) in Fig. 5, which are ingested into the Naval Research Laboratory's 3DVAR scheme [6], reproduced from [29].

One strength of 3DVAR systems is that synthetic or other TC observations from reconnaissance missions can be assimilated easily into the system. The main problem with using 3DVAR systems for TC initialization is that they generally do not have the proper balance constraints for mesoscale phenomena. Most 3DVAR systems have a geostrophic balance condition to relate the mass and wind fields, which is not valid for tropical cyclones and other strongly rotating mesoscale systems, where there exists a nonlinear balance between the mass and wind fields. The improper balance constraint for TCs in 3DVAR systems can result in rapid adjustment during the first few hours of model integration, causing the model vortex to deviate to a state that is very different from the initially ingested synthetic observations. This discrepancy will most likely be carried throughout the forecast period and can cause a large bias for intensity prediction. It has been recently demonstrated how quickly a 3DVAR system can lose the desired TC characteristics [61]. Additionally, it is very hard to use a 3DVAR data assimilation system to adequately capture the secondary circulation correctly, so as to have consistency between the boundary-layer inflow, vertical motion and heating, and outflow.

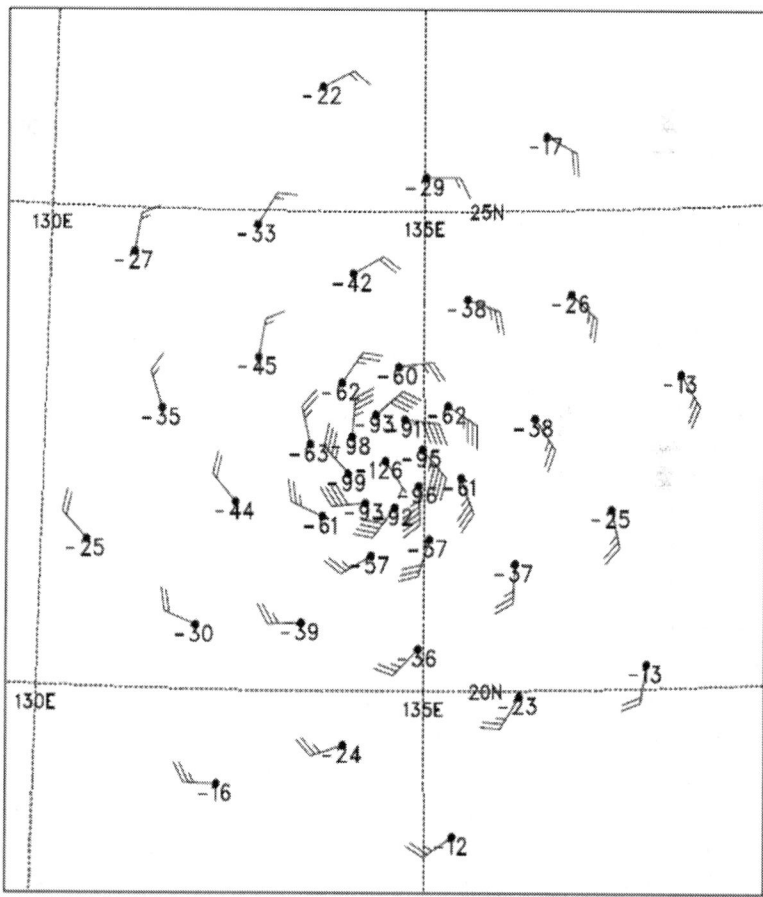

Figure 5: Depiction of near-surface TC synthetic observations for Typhoon Morakot (2009), reproduced from [29]. The synthetic TC observations are blended with all other observations in the 3DVAR data assimilation.

In addition to the synthetic data, dropwindsonde data from aircraft reconnaissance missions may also be included in variational data assimilation systems. Dropwindsondes measure a quasi-vertical profile of the troposphere from where they are launched. A number of studies have shown a positive impact of assimilating dropwindsonde data on TC track [47, 51]. However there can be significant variability on the impact on a case by case basis.

4DVAR Systems

The 4DVAR data assimilation system is a generalization of 3DVAR for assimilating observations that are distributed within a specified time window The goal of 4DVAR is to significantly improve the 3DVAR deficiencies, especially in properly initializing a multi-scale weather system. Compared to 3DVAR, the 4DVAR analyses do not typically show a significant imbalance in the first hours of the forecast. This spin-up process is often associated with the presence of spurious gravity waves that need to be removed by an initialization process (discussed in the next section). A 4DVAR data assimilation system usually requires the development of the tangent linear model and corresponding adjoint system for the forecast model, which are not trivial, in order to iteratively minimize the difference between the first guess fields and the observation. 4DVAR data assimilation systems have been developed for major operation centers for their global prediction system and have led to improvements in forecast skill: ECMWF [40], the Canadian Meterological Centre [14], the U.K. Met Office [41], the Naval Research Laboratory [56], and the Australian Bureau of Meteorology. In some of the 4DVAR systems, synthetic observations are also ingested to improve the TC vortex representation, similar to 3DVAR systems.

An example of an operational TC prediction model that uses a 4DVAR scheme for initialization is ACCESS-TC (Australian Community Climate and Earth System Simulator system for Tropical Cyclones), and a number of other studies have also employed 4DVAR systems for TC initialization [35, 52, 54, 63, 64]. For example, the utility of 4DVAR data assimilation in assimilating irregularly distributed observations in both space and time (such as AMSU-A retrieved temperature and wind fields, as well as the mean sea level pressure (MSLP) information) has been shown by [63]. Using a 72-hour simulation of a land-falling typhoon, they concluded that both the satellite data and the MSLP information could improve the typhoon track forecast, especially for the recurving of the track and landing point. The MM5-4DVAR data assimilation system developed by the Air Force Weather Agency (AFWA) [42] has been employed [62] with a comprehensive satellite products to construct a continuous-coverage, high-resolution TC dataset. Twelve typhoons that occurred over the western Pacific region from May to October 2004 were selected for this reanalysis. The resulting analysis fields show very similar structure of TCs in comparison with satellite observations, demonstrating the capability of 4DVAR in retaining the final structure of the data.

Ensemble Kalman Filter Systems

Another four-dimensional data assimilation system, the ensemble Kalman filter (EnKF), has also been adopted for geophysical models [11, 21]. The Kalman filter is an algorithm which uses a series of measurements observed over time (thus four-dimensional), produces estimates of unknown variables. More formally, the Kalman filter operates recursively on streams of noisy input data to produce a statistically optimal estimate of the underlying system state. The original Kalman Filter assumes that all probability density functions are Gaussian and provides algebraic formulas for the change of the mean and the covariance matrix by the Bayesian update, as well as a formula for advancing the covariance matrix in time provided the system is linear. However, maintaining the covariance matrix is not computationally feasible for high-dimensional systems. For this reason, EnKFs were developed that replace the covariance matrix by the sample covariance computed from the ensemble forecast. The EnKF is now an important data assimilation component of ensemble forecasting. An overview of the work done with the EnKF in the oceanographic and atmospheric sciences can be found in [12].

An intercomparison of an EnKF data assimilation method with the 3D and 4D Variational methods was made using the Weather Research and Forecasting (WRF) model over the contiguous United States during June of 2003 [60]. It is found that 4DVAR has consistently smaller errors than that of 3DVAR for winds and temperature at all forecast lead times except at 60 and 72 h when their forecast errors become comparable in amplitude. The forecast error of the EnKF is comparable to that of the 4DVAR at the 12-36 h lead times, both of which are substantially smaller than that of the 3DVAR, despite the fact that 3DVAR fits the sounding observations much more closely at the analysis time. The advantage of the EnKF becomes even more evident at the 48-72 h lead times.

The EnKF has recently been applied to the TC initialization problem [1, 9, 16, 44, 45, 48, 53, 58, and 59]. The EnKF assimilation of inner-core data, such as airborne Doppler radar winds has shown some promising results with improving the vortex structure and intensity forecasts [1, 57]. In Fig. 6, the performance of an EnKF system for predicting TC intensity is shown for a sample of cases in which airborne Doppler radar data was assimilated reproduced from [57]. As shown in the figure, average intensity errors were reduced by the EnKF assimilation of radar data. [53] Used an ensemble Kalman filter (EnKF) to assimilate center position, velocity of storm motion, and surface axisymmetric wind structure in a high-resolution mesoscale model during the 24-h initialization period to develop a dynamically balanced TC vortex without employing any extra bogus schemes. The surface radial wind profile is constructed by fitting the combined information from both the best-track and the dropwindsonde data available from aircraft surveillance observations, such as the Dropwindsonde Observations for Typhoon Surveillance near the Taiwan Region (DOTSTAR). The subsequent numerical integration shows minor adjustments during early periods, indicating that the analysis fields obtained from this method are dynamically balanced. While the EnKF methods are appealing, due to its ensemble nature, it can be significantly more costly (in a computational sense) than the variational methods.

INITIALIZATION SCHEMES

While the direct insertion and data assimilation techniques can produce estimates of the observed TC, inevitably imbalances will exist after interpolation and analyses procedures. As discussed earlier, the imbalances will typically be greater for the 3DVAR schemes than 4D schemes. The primary purpose of the initialization schemes is to improve the initial dynamic and thermodynamic balances of the TC, so that spurious gravity waves are filtered from the initial condition [5]. In this section, we discuss three widely used initialization schemes:

nonlinear normal mode initialization, digital filters, and dynamic initialization.

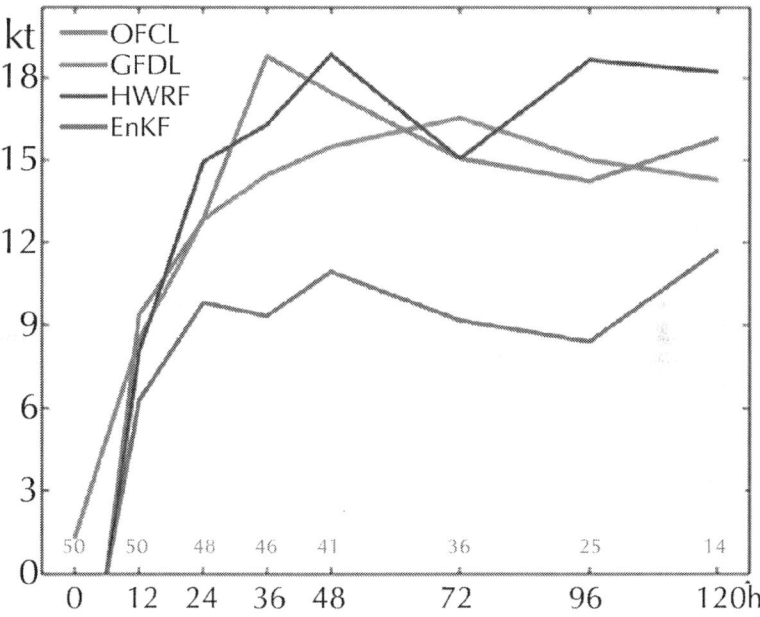

Figure 6: Mean absolute error (ordinate) in the maximum sustained surface wind versus forecast lead time (abscissa) in a homonegeous sample of cases with airborne Doppler radar data during 2008-2010. As shown the EnKF system which assimilates the radar data had a lower average intensity error than the offical National Hurricane Center forecast (OFCL) and other operational hurricane prediction models (GFDL and HWRF). Figure is courtesy of Fuqing Zhang, reproduced from [57] by permission of American Geophysical Union.

Nonlinear Normal Mode Initialization

Since an important goal of initialization to provide a balanced initial state from which minimum spurious gravity activity remains [5], methods have been specifically developed to remove such gravity waves from the initial conditions. An early strategy for removal of high frequency oscillations is the nonlinear normal mode method [2, 33, and 43]. The eigenvalues of the linearized version of the nonlinear forecast model are the normal modes of the system. For a three-dimensional atmospheric model, these normal modes will encompass higher frequency sound and gravity waves, as well as lower frequency Rossby waves. The idea with the normal mode initialization is to project the analysis vector on to the slower modes in order to reduce gravity waves in the initialization.

Digital Filters

Another method to remove high frequency variability is the digital filter. Similar to the electronic analogue, the digital filter performs a mathematical operation on a time signal to reduce or enhance certain aspects of that signal. For atmospheric applications, this is usually accomplished using a filter that has a cutoff frequency, so that waves of a desired frequency can be removed from the analysis [32]. The benefits of the digital filter are that it is a straightforward way to remove waves of a certain frequency without changing the initial condition significantly [22]. The digital filter can be used in both adiabatic and diabatic modes.

Dynamic Initialization

Dynamic initialization (DI) is a short-term integration of the full model before it actually starts the forecast integration to allow the forecast model to handle the spin-up issue. It usually includes two steps: adiabatic backward integration (i.e., to 6 hour) and diabatic forward integration to the initial time. During adiabatic backward integration, the model physics does not contribute to the tendency of the variables so that this process is quasi-reversible (except the effect of numerical diffusion). In the forward integration (i.e., from 6 hour to the actual initial time at zero hour), the model incurs diabatic process with Newtonian relaxation to some chosen variables so that the initial fields are close to the analysis without introducing small model error during the extra integration time. The idea here is, taking TC prediction as an example, that the 3DVAR procedure produced a reasonably accurate initial state, however, imbalances for TCs with their multiple scales will exist and they should be removed prior to the start of model integration. This process also allows for the buildup of the boundary layer and secondary circulation of the TC. The forward

DI can be accomplished by relaxation to any or a combination of the model prognostic variables at the analysis time. Of course, much care should be taken in choosing the proper combination. One commonly adopted DI procedure is to relax to the analysis horizontal momentum during the initialization period. DI can also be enhanced by separately relaxing to the nondivergent and divergent wind components, with different relaxation coefficients [7]. This is useful because the nondivergent winds are better captured by the 3DVAR analysis than the divergent winds, and allows for direct way of including relaxation to the heating profiles (which affect the divergent circulation). Various methods have used to incorporate the diabatic effects into the dynamic initialization procedure. These methods include modifying the humidity vertical profiles due to rain rate assimilation, physical initialization, and dynamic nudging to the satellite observed heating profiles [7, 23, 24, 25, 37, 38, and 39]. As an example of an operational system, the Australian Bureau of Meteorology used a diabatic dynamic initialization scheme in their earlier tropical cyclone prediction system (TC-LAPS). The diabatic, dynamic initialization was used after a high-resolution objective analysis to improve the mass-wind balance of the vortex while building in the heating asymmetries [8].

CONCLUSIONS

This chapter reviewed different methods for initializing TCs in numerical prediction systems. The methods range from simpler direct insertion techniques to more advanced dynamic initialization, and from three-dimensional to four-dimensional data assimilation techniques. The strengths and weaknesses of the different schemes were discussed. The direct insertion techniques take either an analytically specified vortex or a dynamically initialized vortex and insert it into the numerical model analysis. These schemes require removal of the TC vortex in the numerical model first guess or analyzed fields, which is often not at the right location or does not match the observations. The direct insertion schemes are appealing because a vortex can be constructed to match the observations; however, there is no guarantee that when inserting this vortex into the analysis that dynamic and thermodynamic balance will exist. In the data assimilation techniques for TC initialization, synthetic observations matching the observed

TC structure and intensity are created, and a data assimilation system blends these observations with all other observations to generate the analysis. 3DVAR systems are not as well suited for the TC initialization due to its inability to produce a nonlinear balance between the mass and wind fields. 4DVAR and ensemble Kalman filter schemes show some promising results for TC initialization, in particular, in obtaining a better dynamic and thermodynamic balance, and in the case of the EnKF also providing probabilistic information by running an ensemble. Finally, full domain dynamic initialization (adiabatic and diabatic) techniques were discussed. These schemes are advantageous because they are relatively straightforward to implement, and they are able to produce better dynamic and thermodynamically balanced vortices without the development of the four-dimensional data assimilation.

There are a number of significant challenges that remain for TC initialization. First, most TCs lack of observations needed to construct accurate structure for the storms. Only a handful of TCs in the North Atlantic Ocean basin have routine reconnaissance missions. No matter how advanced the initialization system is, it will always be limited by lack or uncertainty in the observations. Secondly, TCs span multiple scales of motion, ranging from turbulence to deep convective updrafts to vortex scale waves (e.g. vortex Rossby waves), to its interaction with the environments and synoptic scale features. While the synoptic scale is largely responsible for TC track, many of these smaller-scale features are important for intensity. These features are transient and unbalanced, leading to initialization challenges. Third, it is difficult to initialize TCs properly in different environments, such as a TC in shear or with dry air wrapping into its core. Finally, if TC intensity largely depends on deep convective evolution, there are inherent limits to predictability.

In spite of these challenges, much progress has been made of the TC initialization front, and there are promising results from the EnKF, 4DVAR and dynamic initialization schemes. The recent trend in data assimilation is to combine the advantages of 4DVAR and the Kalman filter techniques. Considering the threat that TCs will continue to play, efforts must continue to develop enhanced initialization schemes along with the new technologies for data assimilation to better predict track and intensity.

ACKNOWLEDGEMENTS

This research is supported by the Chief of Naval Research through the NRL Base Program, PE 0601153N. The authors thank Jim Doyle and Jon Moskaitis for their comments and assistance.

REFERENCES

1. Altug Aksoy, Sylvie Lorsolo, Tomislava Vukicevic, Kathryn J. Sellwood, Sim D. Aberson, Fuqing Zhang, 2012 The HWRF hurricane ensemble data assimilation system (HEDAS) for high-resolution data: The impact of airborne Doppler radar observations in an OSSE Mon. Wea. Rev. in press

2. F. Baer, J. J. Tribbia, on complete filtering of gravity modes through nonlinear initialization Mon. Wea. Rev. 105 1536 1539 1977

3. Morris A. Bender, Rebecca J. Ross, Robert E. Tuleya, Yoshio M. Kurihara, Improvements in tropical cyclone track and intensity forecasts using the GFDL initialization system Mon. Wea. Rev. 121 2046 2061 1993

4. P. Courtier, E. Andersson, W. Heckley, J. Pailleux, D. Vasiljevic, M. Hamrud, A. Hollingsworth, F. Rabier, M. Fisher, The ECMWF implementation of three-dimensional variational assimilation (3D-Var). Part 1: Formulation Quart. J. Roy. Meteor. Soc. 124 1783 1807 1998

5. Roger Daley, Atmospheric data analysis Cambridge University Press 1991

6. Roger Daley, Edward Barker, NAVDAS: Formulation and diagnostics Mon. Wea. Rev. 129 869 883 2001

7. Noel E. Davidson, Kamal Puri, Tropical prediction using dynamical nudging, satellite-defined convective heat sources, and a cyclone bogus Mon. Wea. Rev. 120 2329 2341 1992

8. Noel E. Davidson, Harry C. Weber, The BMRC high-resolution tropical cyclone prediction system: TC-LAPS Mon. Wea. Rev. 128 1245 1265 2000

9. Jili Dong, Ming Xue, Assimilation of radial velocity and reflectivity data from coastal WSR-88D radars using ensemble Kalman filter for the analysis and forecast of landfalling Hurricane Ike (2008) Quart. J. Roy. Met. Soc. in press

10. Vernon F. Dvorak, Tropical cyclone intensity analysis and forecasting from satellite imagery Mon. Wea. Rev. 103 420 430 1975

11. Geir Evensen, Sequential data assimilation with a nonlinear quasi-geostrophic model using Monte Carlo methods to forecast error statistics J. Geophys. Res. 99 143 162 1994

12. Geir Evensen, The ensemble Kalman filter: theoretical formulation and practical implementation Ocean Dynamics 53 343 367 2003

13. Pierre Gauthier, C. Charette, L. Fillion, P. Koclas, S. Laroche, Implementation of a 3D variational data assimilation system at the Canadian Meteorological Centre. Part I: The global analysis Atmosphere-Oceans 37 103 156 1999

14. Pierre Gauthier, Monique Tanguay, Stephane Laroche, Simon Pellerin, Josee Morneau, Extension of 3DVAR to 4DVAR: Implementation of 4DVAR at the Meteorological Service of Canada Mon. Wea. Rev. 135 233 2354 2007

15. James S. Goerss, Richard A. Jeffries, Assimilation of synthetic tropical cyclone observations into the Navy Operational Global Atmospheric Prediction System Wea. Forecasting 9 557 576 1994

16. Thomas M. Hamill, Jeffrey S. Whitaker, Michael Fiorino, Stanley G. Benjamin, Global ensemble predictions of 2009's tropical cyclones initialized with an ensemble Kalman filter Mon. Wea. Rev. 139 668 688 2011

17. J. T. Heming, J. C. L. Chan, A. M. Radford, A new scheme for the initialisation of tropical cyclones in the UK Meteorological Office global model Meteor. Appl. page DOI: 10.1002/met.5060020211 1995

18. Eric A. Hendricks, Melinda S. Peng, Tim Li, Ge Xuyang, Performance of a dynamic initialization scheme in the Coupled Ocean-Atmosphere Mesoscale Prediction System for Tropical Cyclones (COAMPS-TC) Wea. Forecasting 26 650 663 2011

19. Greg J. Holland, An analytic model of the wind and pressure profiles in hurricanes Mon. Wea. Rev. 108 1212 1218 1980

20. Greg J. Holland, A revised hurricane pressure-wind model Mon. Wea. Rev. 136 3432 3445 2008

21. P. L. Houtemaker, H. L. Mitchell, Data assimilation using an ensemble Kalman filters technique Mon. Wea. Rev. 126 796 811 1998

22. Xiang-Yu Huang, Peter Lynch, Diabatic digital-filtering initialization: Application to the HIRLAM model Mon. Wea. Rev. 121 589 603 1993

23. T. N. Krishnamurti, H. S. Bedi, William Heckley, Kevin Ingles, Reduction in spinup time for evaporation and precipitation in a spectral model Mon. Wea. Rev. 116 907 920 1988

24. T. N. Krishnamurti, Ricardo Correa-Torres, Greg Rohaly, Darlene Oosterhof, Naomi Surgi, Physical initialization and hurricane ensemble forecasts Wea. Forecasting 12 503 514 1997

25. T. N. Krishnamurti, Wei Han, Bhaskar Jha, H.S. Bedi, Numerical prediction of Hurricane Opal Mon. Wea. Rev. 126 1347 1363 1998

26. Yoshio M. Kurihara, Morris A. Bender, Rebecca J. Ross, An initialization scheme of hurricane models by vortex specification Mon. Wea. Rev. 121 2030 2045 1993

27. Yoshio M. Kurihara, Morris A. Bender, Robert E. Tuleya, Rebecca J. Ross, Improvements in the GFDL Hurricane Prediction System Mon. Wea. Rev. 123 2791 2801 1995

28. Lance M. Leslie, G. J. Holland, on the bogussing of tropical cyclones in numerical models: A comparison of vortex profiles Meteorol. Atmos. Phys. 56 101 110 1995

29. C. S. Liou, Keith D. Sashegyi, On the initialization of tropical cyclones with a three-dimensional variational analysis Natural Hazards 2012 631375 1391

30. A. C. Lorenc, S. P. Ballard, R. S. Bell, N. B. Ingleby, P. L. F. Andrews, D. M. Barker, J. R. Bray, A. M. Clayton, T. Dalby, D. Li, T. J. Payne, F. W. Saunders, The Met. Office global three-dimensional variational data assimilation scheme Quart. J. Roy. Meteor. Soc. 126 2991 3012 2000

31. A. Lorenz, Analysis methods for numerical weather prediction Quart. J. Roy. Meteor. Soc. 112 1177 1194 1986

32. Peter Lynch, Xiang-Yu Huang, Initialization of the HIRLAM model using a digital filter Mon. Wea. Rev. 120 1019 1034 1992

33. B. Machenhauer, on the dynamics of gravity oscillations in a shallow water model, with applications to normal mode initialisation Beitr. Phys. Atmos. 50 253 271 1977

34. Makut B. Mathur, The National Meteorological Center"s quasi-Lagrangian model for hurricane prediction Mon. Wea. Rev. 119 1419 1447 1991

35. Kyungjeen Park, X. Zou, toward developing an objective 4DVAR BDA scheme for hurricane initialization based on TPC observered parameters Mon. Wea. Rev. 132 2054 2069 2004

36. David F. Parrish, John C. Derber, The National Meteorological Center"s spectral statistical-interpolation analysis system Mon. Wea. Rev. 120 1747 1763 1992

37. Melinda S. Peng, Simon W. Chang, Impacts of SSM/I retrieved rainfall rates on numerical prediction of a tropical cyclone Mon. Wea. Rev. 124 1181 1198 1996

38. Melinda S. Peng, B. F. Jeng, C. P. Chang, Forecast of typhoon motion in the vicinity of Taiwan during 1989-90 using a dynamical model Wea. Forecasting 8 309 325 1993

39. K. Puri, N. E. Davidson, The use of infrared satellite cloud imagery data as proxy data for moisture and diabatic heating in data assimilation Mon. Wea. Rev. 120 2329 2341 1992

40. F. Rabier, H. Jarvinen, E. Klinker, J.-F. Mahfouf, A. Simmons, The ECMWF operational implementation of four-dimensional variational assimilation. I: Experimental results with simplified physics Quart. J. Roy. Meteor. Soc. 126 1143 1170 2000

41. F. Rawlins, S. P. Ballard, K. J. Bovis, A. M. Clayton, D. Li, G.W. Inverarity, A.C. Lorenc, T. J. Payne, The Met Office global four-dimensional variational data assimilation scheme Quart. J. Roy. Meteor. Soc. 133 347 362 2006

42. F. H. Ruggiero, J. Michalakes, T. Nehrkorn, G. D. Modica, X. Zou, Development of a new distributed-memory MM5 adjoint J. Atmos. Ocean Tech. 23 424 436 2006

43. C. Temperton, Implicit normal mode initialization Mon. Wea. Rev. 116 1013 1031 1988

44. Ryan D. Torn, Performance of a mesoscale ensemble Kalman filter (EnKF) during the NOAA high-resolution hurricane test Mon. Wea. Rev. 138 4375 4392 2010

45. Ryan D. Torn, Greg J. Hakim, Ensemble data assimilation applied to RAINEX observations of Hurricane Katrina (2005) Mon. Wea. Rev. 137 2817 2829 2009

46. Yuqing Wang, on the bogusing of tropical cyclones in numerical models: The influence of vertical tilt Meteorol. Atmos. Phys. 65 153 170 1998

47. Martin Weissmann, Florian Harnisch, Wu Chun-Chieh, Po-Hsiung Lin, Yoichiro Ohta, Koji Yamashita, Yeon-Hee Kim, Eun-Hee Jeon, Tetsuo Nakazawa, Sim Aberson, The influence of assimilating dropsonde data on typhoon track and midlatitude forecasts Mon. Wea. Rev. 139 908 920 2011

48. Yonghui Weng, Fuqing Zhang, Assimilating airborne Doppler radar observations with an ensemble Kalman filter for convection-permitting hurricane initialization and prediction: Katrina (2005) Mon. Wea. Rev. 140 841 859 2012

49. Hugh E. Willoughby, Gradient balance in tropical cyclones J. Atmos. Sci. 47 265 274 1990

50. Henry R. Winterbottom, Eric P. Chassignet, A vortex isolation and removal algorithm for numerical weather prediction model tropical cyclone applications J. Adv. Model. Earth. Sys. 3 M11003 8 2011

51. Wu Chun-Chieh, Kun-Hsuan Chou, Po-Hsiung Lin, Sim D. Aberson, Melinda S. Peng, Tetsuo Nakazawa, The impact of dropwindsonde data on typhoon track forecasts in DOTSTAR Wea. Forecasting 22 1157 1176 2007

52. Wu Chun-Chieh, Kun-Hsuan Chou, Yuqing Wang, Ying-Hwa Kuo, Tropical cyclone initialization and prediction based on four-dimensional variational data assimilation J. Atmos. Sci. 63 2383 2395 2006

53. Wu Chun-Chieh, Guo-Yuan Lien, Jan-Huey Chen, Fuqing Zhang, Assimilation of tropical cyclone track and structure based on the ensemble Kalman filter (EnKF) J. Atmos. Sci. 67 3806 3822 2007

54. Pu Zhao Xia, Scott A. Braun, Evaluation of bogus vortex techniques with four-dimensional variational data assimilation Mon. Wea. Rev. 129 2023 2039 2001

55. Qingnong Xiao, Ying-Hwa Kuo, Ying Zhang, Dale M. Barker, Duk-Jin Won, A tropical cyclone bogus data assimilation scheme in the MM5 3D-Var system and numerical experiments with Typhoon Rusa (2002) near landfall J. Meteor. Soc. Japan 84 671 689 2006

56. Xu Liang, Tom Rosmond, Roger Daley, Development of NAVDAS-AR: Formulation and initial tests of the linear problem Tellus 57A 546 559 2005

57. Fuqing Zhang, Yonghui Weng, John F. Gamache, Frank D. Marks, Performance of convection-permitting hurricane initialization and prediction during 2008-2010 with ensemble data assimilation of inner-core airborne Doppler radar observations Geophys. Res. Lett. 38 L15810 2011

58. Fuqing Zhang, Yonghui Weng, Ying-Hwa Kuo, Jeffrey S. Whitaker, Baoguo Xie, Predicting Typhoon Morakot''s catastrophic rainfall with a convection-permitting mesoscale ensemble system Wea. Forecasting 25 1861 1825 2010

59. Fuqing Zhang, Yonghui Weng, Jason A. Sippel, Zhiyong Meng, Craig H. Bishop, Cloud-resolving hurricane initialization and prediction through assimilation of Doppler radar observations with an ensemble Kalman filter Mon. Wea. Rev. 137 2105 2125 2009

60. Meng Zhang, Fuqing Zhang, Xiang-Yu Huang, Xin Zhang, Intercomparison of an ensemble Kalman filter with three- and four-dimensional variational data assimilation methods in a limited-area model over the month of June 2003 Mon. Wea. Rev. 139 566 572 2011

61. Shengjun Zhang, Tim Li, Ge Xuyang, Melinda S. Peng, Ning Pan, A 3DVar-based dynamical initialization scheme for tropical cyclone predictions Wea. Forecasting 27 473 483 2012

62. X. Zhang, T. Li, F. Weng, C. C. Wu, L. Xu, Reanalysis of western Pacific typhoons in 2004 with multi-satellite observations Meteorol. Atmos. Phys. 97 3 18 2007

63. Y. Zhao, B. Wang, Z. Ji, X. Liang, G. Deng, X. Zhang, Improved track forecasting of a typhoon reaching landfall from four-dimensional variational data assimilation of AMSU-A retrieved data J. Geophys. Res. 110 D14101 2005

64. Ying Zhao, Bin Wang, Juanjuan Liu, A DRP-4DVar data assimilation scheme for typhoon initialization using sea level pressure data Mon. Wea. Rev. 140 1191 1203 2012

65. X. Zou, Q. Xiao, Studies on the initialization and simulation of a mature hurricane using a variational bogus data assimilation scheme J. Atmos. Sci. 57 836 860 2000

Effects of Urban Configuration on Human Thermal Conditions in a Typical Tropical African Coastal City

Emmanuel Lubango Ndetto[1,2] and Andreas Matzarakis[1]

[1]Chair of Meteorology and Climatology, Albert-Ludwigs University of Freiburg, Werthmannstrasse 10, 79085 Freiburg, Germany
[2]Department of Physical Sciences, Sokoine University of Agriculture, Morogoro, Tanzania

ABSTRACT

A long-term simulation of urban climate was done using the easily available long-term meteorological data from a nearby synoptic station in a tropical coastal city of Dar es Salaam, Tanzania. The study aimed at determining the effects of buildings' height and street orientations on human thermal conditions at pedestrian level. The urban configuration was represented by a typical urban street and a small urban park near the seaside. The simulations were conducted in the microscale applied climate model of RayMan, and results were interpreted in terms of the thermal comfort parameters of mean radiant (T_{mrt}) and physiologically equivalent (PET) temperatures. PET values, high as 34°C, are observed to prevail during the afternoons especially in the east-west oriented streets, and buildings' height of 5 m has less effect on the thermal comfort. The optimal reduction of T_{mrt} and PET values for pedestrians was observed on the nearly north-south reoriented streets and with increased buildings' height especially close to 100 m. Likewise, buildings close to the park enhance comfort conditions in the park through additional shadow. The study provides design implications and management of open spaces like urban parks in cities for the sake of improving thermal comfort conditions for pedestrians.

INTRODUCTION

Dar es Salaam is a hot, humid tropical city situated along the Tanzanian coast on the Western Indian Ocean coast. It enjoys the sea-land breezes system which is observed to be more organised during the months of March, April, September, and October [1, 2]. On the other hand, sultriness is not uncommon particularly during the December–February (DJF) season. Such sultry conditions bring an undesirable thermal discomfort especially to pedestrians and street vendors. Most biometeorological studies done elsewhere have revealed the effects of street orientation and height to width ratio on the variation of local thermal comfort at an urban canyon level [3–6].

Although the human thermal comfort is influenced by four meteorological parameters of air temperature, humidity, wind speed, and radiation flux (usually quantified in terms of mean radiant temperature), it is the wind speed and mean radiant temperature which can be significantly varied by changing the street orientation and height to width ratio [7]. A past urban climate research in Dar es Salaam has suggested that the observed sea-land breeze effect could improve the thermal comfort at high ground places in the city like at the university of Dar es Salaam campus which is about 12 km to the west from the city centre [1]. Moreover, various methods have been applied elsewhere in order to deepen the understanding of the effect of street geometry and orientation on urban thermal comfort especially at pedestrian level. For instance, from the field-measurements in an east-west urban canyon during cloudless summer weather in 2003, it was found that thermal stress is mostly attributable to solar exposure, and the study suggested further field investigations to be done in other locations and climatic conditions in order to verify the generality of such thermal observations in street canyons [3].

In Fez, Morocco, measurements were carried out during the hot summer and cool winter seasons where the deep canyons were found considerably cooler in day than the shallow canyons [8]. Effects of urban configuration on urban thermal climate have also been recently explored through series of field measurements in many cities including Tokyo, Japan [9] Athens, Greece [4] Constantine [5], Colombo [10, 11]. In some of the cities situated within the coastal environments, influence of sea breezes on urban thermal climate was also investigated [10, 12, 13].

Likewise, numerical simulation also offers a valuable method to understanding the effects of urban geometry and street orientation. In Ghardaia, Algeria, numerical simulations were done using the three-dimensional numerical model, ENVI-met, for a typical summer day, and results showed contrasting patterns of thermal comfort between shallow and deep urban streets as well as between various orientations [6]. In studying the impact of street geometry on ambient temperatures and on daytime pedestrian comfort levels, two approaches involving field measurements and urban climate simulations using the ENVI-met were carried out in downtown of Curitiba, Brazil [14]. Having observed that maximum daily temperature within street canyons in Colombo, Sri Lanka, decreases with increasing height to width (H/W) ratio and that sea breezes exert a cooling effect, simulations done in ENVI-met model were initiated by data obtained from a synoptic weather station located at the airport (approximately 24 km north of the measurement location) for the purpose of understanding the effect of different urban design options on air and surface temperatures, as well as on outdoor thermal comfort [15]. Similarly, long-term thermal comfort at the University Campus in Taiwan was predicted by using long-term meteorological data collected from a nearby station, and simulations were performed in RayMan model [7]. This suggests that with modern advances in urban climatology especially through continuous improvement on the available microscale models, assessment of thermal conditions in urban places could then be sufficiently studied where simulations could deliver helpful results. Besides, most field studies examining outdoor thermal comfort merely clarify characteristics measured on a particular day; hence, such studies may not represent annual thermal conditions accurately.

Due to its coastal location and proximity to the equator, Dar es Salaam seems to be a perfect location of understanding the effects of urban geometry on human thermal comfort at pedestrian level in the low latitudes. Long-term analysis of urban thermal climate in Dar es Salaam is therefore of paramount importance in understanding the tropical urban climate in a typical African city.

The current study, therefore aimed at quantifying the effects of the buildings' height and orientation on radiation fluxes in a typical urban canyon and an urban park in a tropical African city. In order to optimise thermal comfort through modified urban configurations, human thermal index of physiologically equivalent temperature (PET) was used in visualizing the changes in terms of thermal comfort classes after every modification. Simulations were performed using the microscale model of RayMan [16, 17]. The model is reputed to be easy to use. Recent research done in a midlatitude city found that RayMan can even deliver more accurate results at high sun elevations [18], and this underpins it as an ideal microscale model to study urban climate in the low latitude cities including Dar es Salaam where, as elsewhere in the world, the afternoons are usually the most thermal stressful times [1, 2]. Simulating the optimal configurations for thermal comfort in Dar es Salaam is of utmost importance in advancing the science of urban climate in low latitudes and urban planning of cities in developing countries. Dar es Salaam is an important economic centre in eastern Africa due to its harbour services, and it has recently been observed to expand rapidly. Better understanding of its urban climate is significant for its future urban planning and well-being of the increasing population.

METHODS

Dar es Salaam (Figure 1(a)) is located at 6°51'S, 39°18'E along the south western coast of the Indian ocean, covering an area of 1350 km² of which about 1000 km² is a land area [19, 20]. Dar es Salaam is generally a lowland area with its altitude ranging from the sea level at the coast to an approximately 250 m in the south-west along the Pugu hills situated about 25 km from the city centre [21]. The climate is typically hot-humid, referred to as tropical wet/dry climate (Aw) according to the Köppen classification system [22, 23]. The climate is mainly influenced by the northeast monsoon which prevails from March to October and the southeast monsoon between October and March [19]. This is in response to the passage of the Intertropical convergence zone (ITCZ). Local winds are generally high up to 13 m/s during the afternoons particularly from August to November.

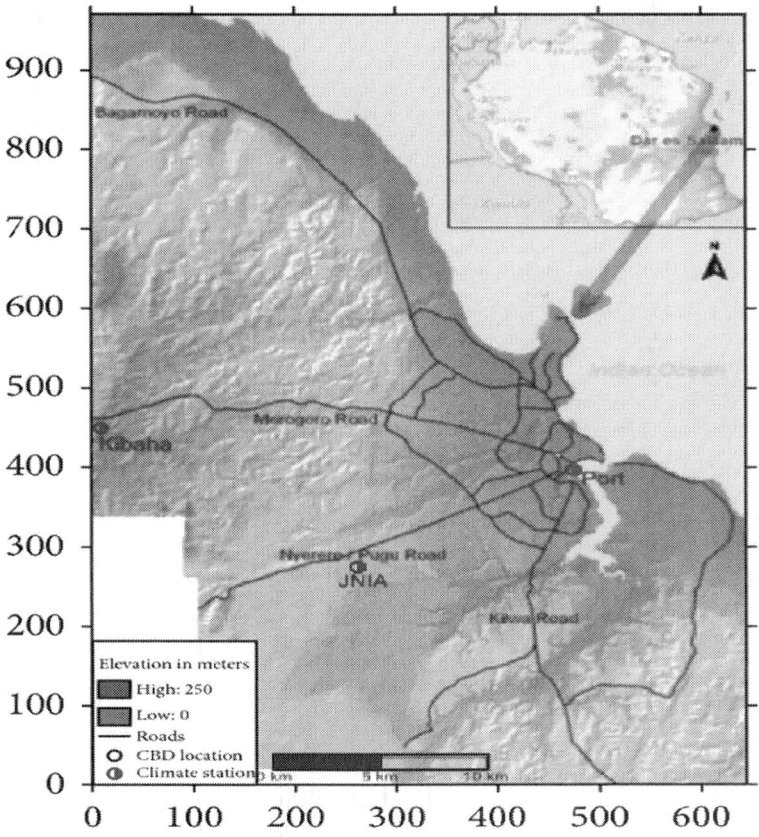

(a)

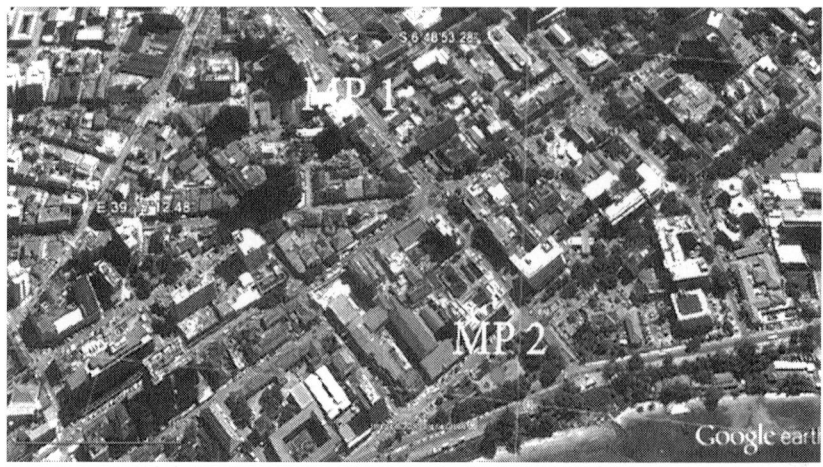

(b)

Figure 1: (a) Location of measurement stations and area of study. Source: modified from Hill and Lindner (2010). (b) A satellite picture as obtained from Google Earth, locating the simulation points, MP1 (typical urban street) and MP2 (urban park), in a part of Dar es Salaam city, Tanzania.

Usually, the relative humidity in Dar es Salaam ranges between 67 and 96% in a year, and the annual rainfall is about 1050 mm with peaks in April and December. April is however the wettest month, and the rain seasons are usually described as short rains (October–December season) with an average of 75 to 100 mm and long rains (March–May season) with a monthly average of 150 to 300 mm of rainfall [24]. The mean annual air temperature is about 30°C with a slight seasonal change due to its proximity to the equator. The mean daily sunshine duration is about 10–12 hours.

The study area comprised of two urban resort places (Figure 1(b)). The first area (MP1) is along the typical street (width of 24.5 m) in the downtown area of Dar es Salaam. One side of the street there is a tall building of 21 storeys (about 93 m high), and the other side is a waiting area for boarding public transport. The second area of simulation (MP2) is on a small urban park (locally called as a Posta garden) overlooking the sea to the south. It also serves as the bus waiting and recreation place. The street is originally orientated in a northwest-southeast direction (i.e., about 45° from the north in an anticlockwise direction) while the park takes the triangle-like shape measuring 156 m × 140 m × 60 m with its hypotenuse facing the sea. There are few trees of mainly Ashoka type along the street whereas the park has many trees with different species, but notably the leaf trees of Neem species, grass, and a concrete floor at the centre. Although the park overlooks the sea, the presence of two lines of trees may act as an obstacle to the sea breezes at low levels or near the ground level. Simulations were therefore done by rotating the orientation every 15° and by increasing the buildings' height by 5 m up to 50 m, and 100 m was taken as the maximum height of the current highest building in the city.

The input meteorological data used for the long-term simulation were based on the available meteorological data collected at a synoptic station located at the Julius Nyerere International Airport (JNIA) (see Figure 1(a)). These data were from the year 2001 to 2012. The airport station usually does meteorological observations at every synoptic hour. However, the available data used in this study had a three-hour frequency period, with observation times at 0 LST (local standard time), 3 LST, 6 LST, 9 LST, 12 LST, 15 LST, 18 LST, and 21 LST. These include air temperature, wind speed, relative humidity, and amount of cloud cover. Cloud cover was used to estimate the radiation flux as there are only daily measurements of radiation at the synoptic station. The airport datasets were used on the assumption that they form a continuous long-term dataset as required in long-term climatic analysis, unlike the datasets from the other two stations. The urban effect at MP1 and MP2 locations was considered to affect the wind speed which was then estimated from the readings at the airport station using equation provided in [25] and applied in [26]:

$$ws_{1.1} = ws_h \left(\frac{1.1}{h} \right)^{\propto}, \qquad \propto = 0.12z_0 + 0.18, \tag{1}$$

Where ws_h is the wind speed (ms^1) at a height of h (10 m), $\propto$ is an empirical exponent that depends on the surface roughness, and z_0 is the roughness length. Due to slight differences in terrain features at MP1 (a densely built-up city area) and MP2 (partly wooded areas with buildings and open to sea), z_0 were taken to be 1.5 and 1.3, respectively, though both areas are within the inner city.

Additional meteorological datasets of air temperature (including maximum and minimum) and relative humidity were also obtained from other two stations located at the harbour (Port) and at the Kibaha Sugarcane Research Institute (Kibaha) (see Figure 1(a) and Table 1). The air temperature datasets from the two stations were from the year 2001 to 2011, whereas relative humidity data spanned for the period March–September 2005 was only collected at the Port station. These additional datasets were compared with the corresponding datasets from the airport station in order to establish the rationale of using long-term synoptic data to simulate urban bioclimatic conditions. Comparison between the weather elements from the three weather stations was examined using the rank correlation of Kendall tau correlation at a 0.05 significance level. The Kendall tau is a bivariate measure of correlation/ association usually used for the rank-order data [27].

Table 1: Climate measuring stations in Dar es Salaam

Meteorological station	Grid reference	Altitude (m)	Approximate distance to area of study (km)
JNIA	06°52 S, 39°12 E	53.0	11.2
Port	06°50 S, 39°18 E	18.0	0.8
Kibaha	06°50 S, 38°58 E	167.0	35.2

Source: Tanzania meteorological agency.

The simulated bioclimatic conditions at pedestrian level were interpreted using PET and T_{mrt}. PET evaluates the thermal conditions in a human physiological manner and uses a commonly known unit of degree Celsius. PET is defined as the air temperature at which the human energy budget for the assumed indoor conditions is balanced by the same skin temperature and sweat rate as under the actual complex outdoor conditions to be assessed [28–30]. On the other hand, the mean radiant temperature is defined as the uniform temperature of an imaginary enclosure in which the radiant heat transfer from the human body equals the radiant heat transfer in the actual nonuniform enclosure [31]. T_{mrt} is one of the meteorological parameters that govern human thermal energy balance hence affecting the human thermal comfort [32].

As the aim of the study was to determine the optimal urban configuration for the urban thermal comfort in Dar es Salaam at pedestrian level, long-term simulation would give reliable and comprehensive information as opposed to short field measurements. Field studies examine outdoor thermal comfort that merely expound characteristics measured on a particular day and would not accurately represent the annual thermal conditions. Besides, the lack of urban meteorological stations with continuous measurements in many tropical cities including Dar es Salaam lead to many urban climate studies in such cities to rely on the easily available long-term meteorological data from nearby stations [7, 15, 33–35]. For instance, climatic changes in temperature conditions in Nairobi city were investigated using data collected from weather stations situated about 4 kilometres from the city centre [33], and in Taiwan, long-term thermal environment was simulated using meteorological data of a 10-year period collected from a nearby station [7]. However, altitude is acknowledged as a determinant factor for the spatial variation of weather and climate, and winds are experienced to decrease in urban areas [36]. Taking into consideration such observations, variation of weather in Dar es Salaam due to altitude could be assumed insignificant as the synoptic station lies at an altitude of 53 m above sea level while the city centre is at about 11 m above sea level. Theoretically, air temperature varies by 0.6°C for every 100 m high. Furthermore, mean climatic conditions of Dar es Salaam are defined using data from the airport station.

The assumptions that guided the use of the available synoptic meteorological data as input in the simulations thus base on the fact that Dar es Salaam lies within the coastal lowland terrain, and a weather change due to altitude is insignificant within few horizontal distances. A robust and plausible long-term urban thermal characteristic can be attained by using long-term and continuous meteorological data. Although differences exist in some atmospheric parameters between the simulation points in the city and the synoptic weather station, the attributed changes in thermal indices generated by the simulation model are relatively small and can be neglected. Therefore, the simulated results provide a precise representation of the thermal conditions within the city environment.

RESULTS AND DISCUSSION

Local Urban Micrometeorological Variations

The interurban variation of local meteorological conditions in Dar es Salaam was investigated in order to determine to what extent does the local meteorological conditions change in spatial and temporal scales. Analysis was done only in terms of relative humidity and air temperature due to availability of these meteorological parameters in the three weather stations. Spatial variability in cloud cover could be understood to be within the allowable limits as the observation at the synoptic station is considered to span a horizontal distance of nearly 30 km from the observation point for the horizontal visibility and consequently cloud cover. This analysis was of course considered necessary and of utmost importance to understand the variability of the local weather, and hence, how reliable can the weather parameters at the synoptic weather station be considered representative of the city's climate. Further, variability analysis of weather parameters was also considered as an important step to establish a rationale of using the meteorological data from the synoptic station as the representative input parameter in the long-term urban climate simulation.

The results, therefore, suggested that the variation of the local micrometeorological conditions in Dar es Salaam could be highly influenced by its proximity to the sea. Places close to the sea could be slightly warmer than those far away from the sea as suggested by the analysis of air temperature from the three stations. Figure2 depicts the long-term mean monthly air temperature differences (ΔT_a) where air temperature at the Port station could be on average 1.5°C or higher than the readings at the airport and Kibaha stations. The monthly mean air temperature differences between JNIA and Port varied from −0.4°C in February to −1.7°C in June and between Port and Kibaha ranged from 0.3°C in October to 1.8°C in May. Between JNIA and Kibaha, the intraurban monthly mean air temperature differences were up to 0.6°C from December to May, warm at JNIA and cold by nearly 0.5°C during June–November season. The intraurban air temperature differences were observed to be more consistent in terms of the mean daily air temperatures in a sense that it is warmer near the seaside (Port station) than inland of the coast (see Figure 2).

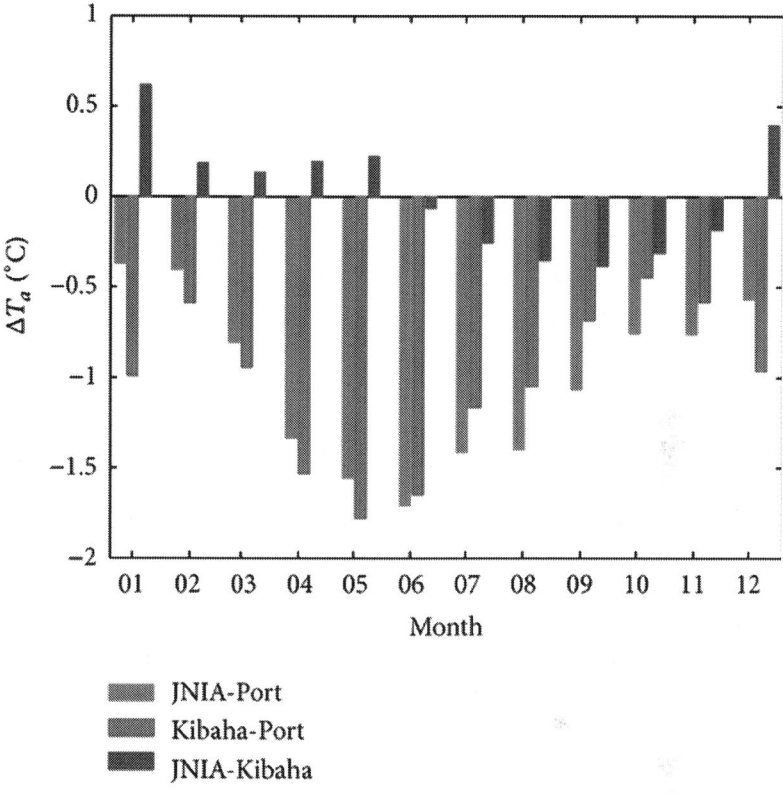

Figure 2: Intraurban air temperature differences as depicted by the long-term mean monthly differences between the three weather stations.

The relationship between the evolutions of temperature profiles among the three measuring stations as determined by the Kendall tau coefficient was found to be as low as 0.25 and 0.27 between the maximum air temperature at the Port and that of other stations, respectively (Table 2). On the contrary, the correlation coefficients for the mean and minimum air temperature between the three stations was high as 0.6 suggesting a similar evolution of temperature at the stations. Usually, the maximum temperature is reached at 15 LST where the surface receives much of the solar insolation. The low correlation coefficients between the seaside station and the inland stations during the maximum temperature could be explained by the differences in terrain at the measuring stations. While Port is near to the sea, terrain features at the inland stations is mostly consisted by green vegetation, usually short grass at the airport (JNIA) and scattered trees at Kibaha. It could be suggested that the differences in thermal capacity between the mostly vegetated terrain station and the station near the large water body is pronounced during the daytime. Other studies also found a trend towards a greater difference between annual mean of daily maximum and that of minimum temperatures from the coast to inland stations [37].

Table 2: Kendall's tau correlation coefficients at a significant level of 0.05 for intra-urban temperature and relative humidity differences between the three stations

	JNIA-Port	Kibaha-Port	JNIA-Kibaha
Ta maximum	0.25	0.27	0.72
Ta minimum	0.87	0.86	0.87
Ta mean	0.70	0.60	0.67
RH mean	0.49	—	—
RH at 15 LST	0.43	—	—

Relative humidity is always high along the eastern Africa coast, but our analysis for relative humidity was limited by availability of data in Dar es Salaam area. Although two of the stations (JNIA and Port) can do observation of relative humidity, available data at Port station were limited to six months of the year 2005. This cannot give a conclusive spatial distribution of relative humidity in Dar es Salaam. With such shortcomings, the current analysis, however, surprisingly indicated that relative humidity is slightly higher at the airport station than at a seaside station especially from April to July (Figure 3). The significant correlation coefficients between readings of relative humidity at the port and the airport stations for the mean monthly and at 15 LST (time of lowest relative humidity) are given in Table 2. The relative humidity climatology of Dar es Salaam as observed from previous studies indicates that relative humidity along the coast usually reach it daily maximum during the night at 90–95% throughout the year, and greater variations are experienced during the daily minimum period at 15 LST [37]. While the seaside station is very close to the city and despite the presence of a large water body, relative humidity readings can as well be influenced by the impervious surface of the urban on the other side. The airport station is well within a short-grass vegetation area. Furthermore, the months of April and May are usually rainy months in Dar es Salaam, and perhaps the proportion difference in terrain features and rain events can influence high humidity in many places not necessarily being close to the sea. However, a good picture of intraurban relative humidity variation in Dar es Salaam could be comprehensive with more datasets with good spatial resolution.

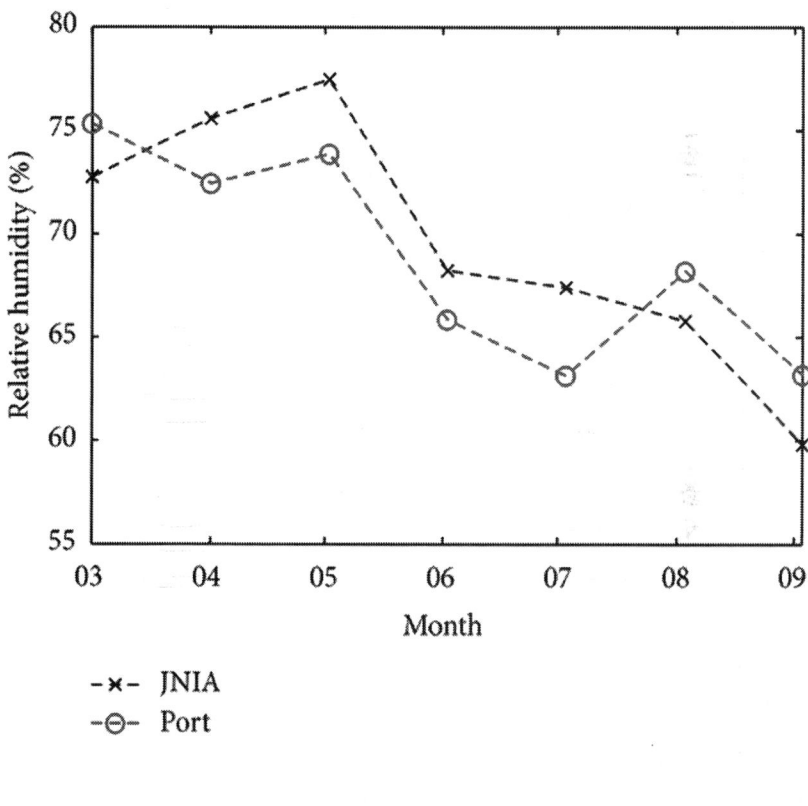

(a)

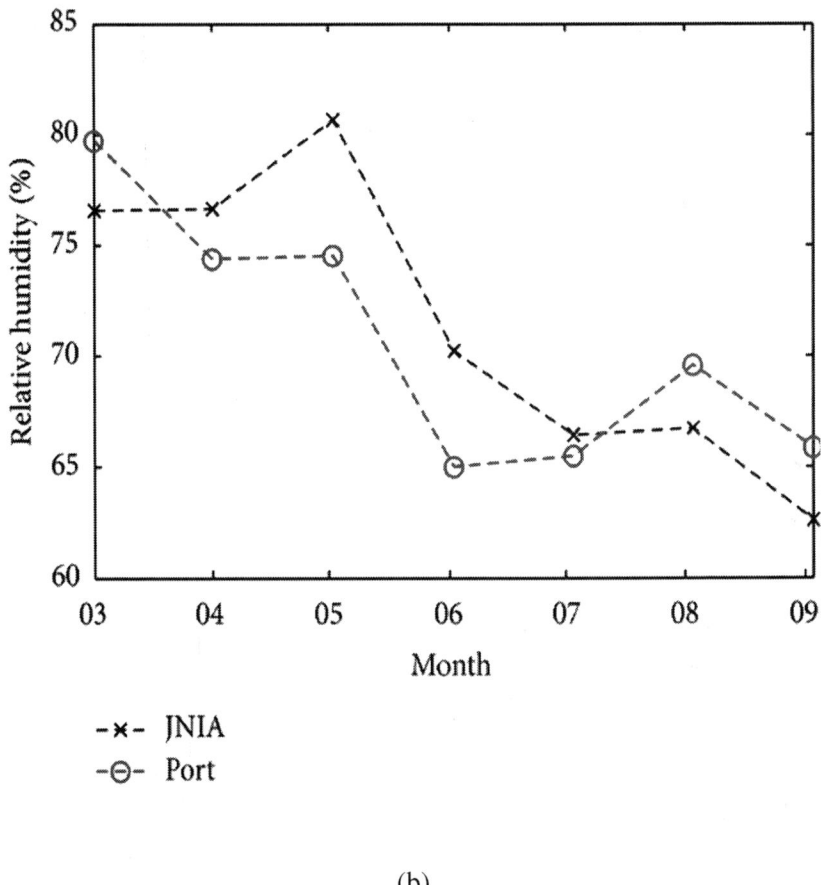

(b)

Figure 3: Mean monthly relative humidity at JNIA and Port (a) and its corresponding mean relative humidity at 15 LST (lowest relative humidity for the day) (b), for the period March–September 2005.

Simulation Results in the Street

The simulations of varying the street orientation and height of buildings were done at two popular urban places in Dar es Salaam for a purpose of quantifying the effects of street orientation and buildings' height on the human thermal comfort at pedestrian level. This section therefore describes the results of the simulation performed at the street and at the urban park.

In its original orientation of a northwest-southeast direction and settings of one side with tall buildings, pedestrians at the street can experience thermal stress during the afternoons. The temporal distribution of as depicted in Figure 4 indicates high values of about 45°C occurring at around 12 LST from the end of November through the beginning of January as well in March. When reoriented in the west-east direction, the highest values of T_{mrt} above 45°C became prevalent at 15 LST in October and from late January to early March while in the north-south reorientation, high values of T_{mrt} of more than 45°C prevailed at noon time from late November to early March.

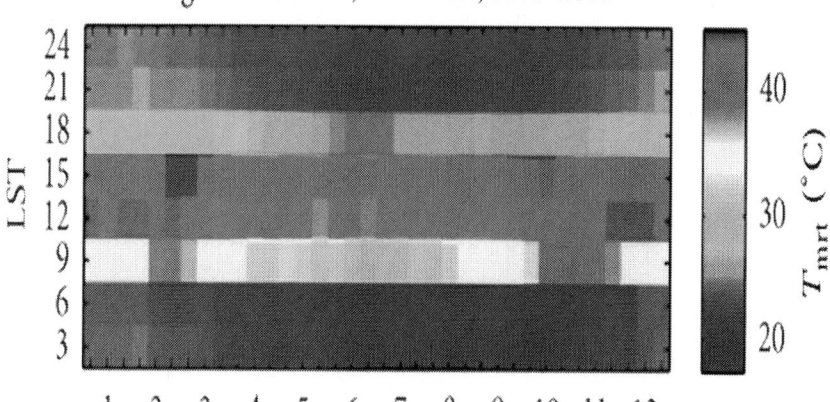

(a)

East-west orientation, 2001–2012

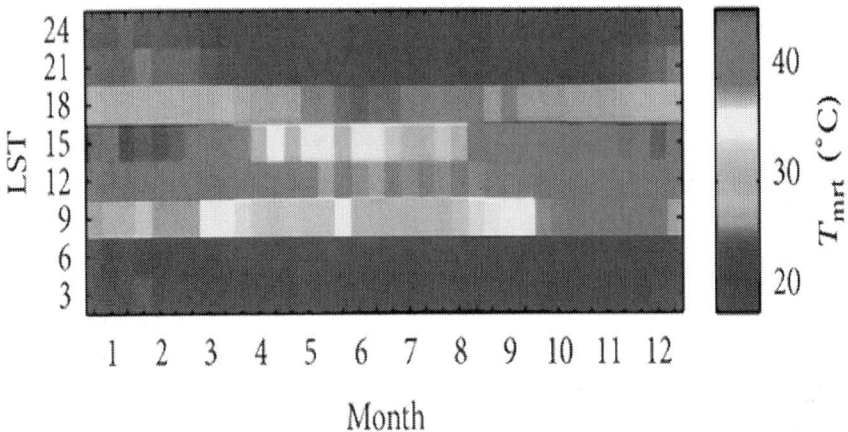

(b)

North-south orientation, 2001–2012

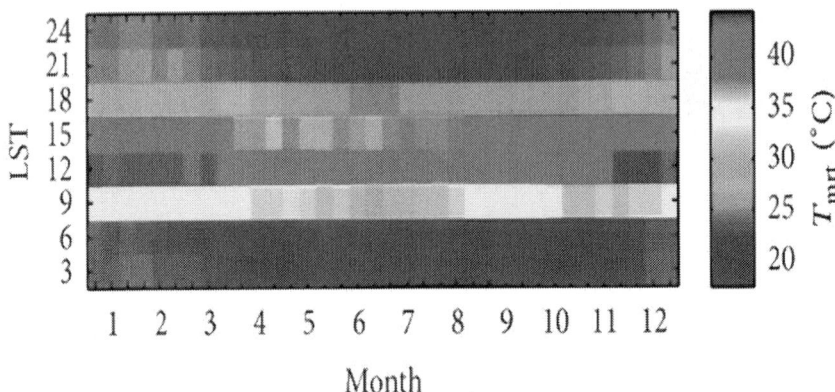

(c)

Figure 4: The temporal distribution of T_{mrt} at the urban street in its origin orientation of a northwest-southeast direction (a), east-west orientation (225°) (b) and north-south orientation (315°) (c).

The corresponding situation in terms of human thermal comfort was analysed using PET index and depicted in Figure 5. The temporal distribution then indicated that high value of PET as 34°C was prevalent between 12 and 15 LST in the months of February, March, November, and December. Further, in an east-west reorientation, high PET values of about 34°C could frequently occur between 12 and 15 LST in March and November whereas in the north-south reorientation similar high values would be experienced in February, March, November, and December in the same hours of the day. The same period was also indentified in the earlier studies as the most thermal stress period in Dar es Salaam even without considering urban obstacles in calculating the thermal indices [1, 2]. Thus, in order to visualize the thermal comfort characteristics in Dar es Salaam, an idealized urban canyon was considered in further analysis.

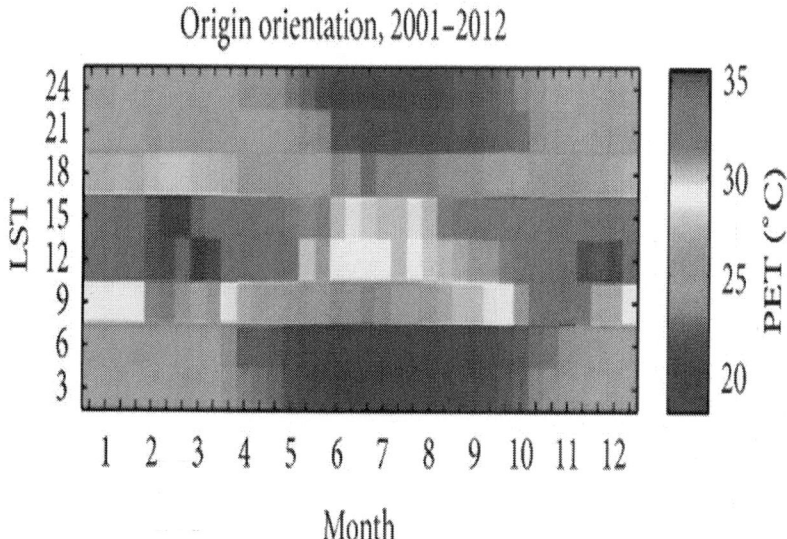

(a)

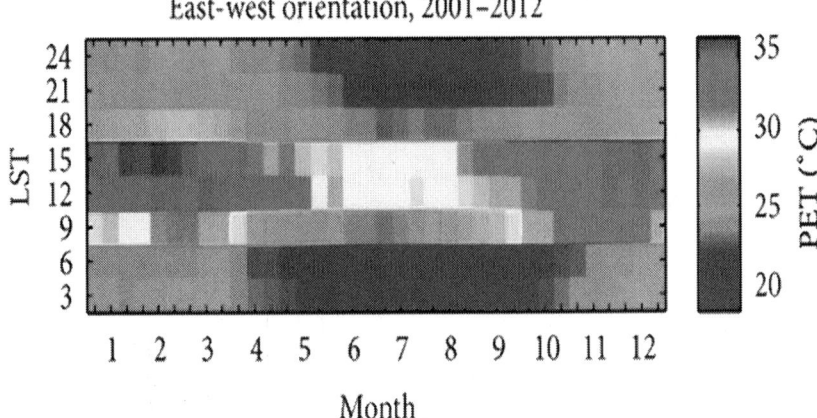

(b)

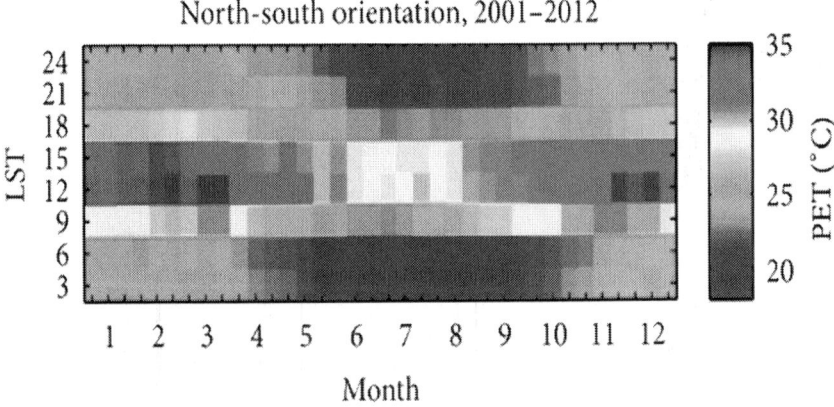

(c)

Figure 5: The temporal distribution of thermal comfort conditions in terms of PET index at the urban street in its origin orientation of a northwest-southeast direction (a), east-west orientation (225°) (b), and north-south orientation (315°) (c), suggesting a high prevalence of human thermal stress between 12 and 15 LST particularly from late October to early April.

An idealized urban canyon with equal heights of buildings on each side was then formulated based on the settings of the original street, and simulations were done by varying the orientation and height of buildings. It is important to note that the new city master plan strategize on densification of the central business district (CBD) of Dar es Salaam through erecting buildings of more than 10 storeys while leaving most of the streets' widths unchanged. Results in this setting are described in Figures 6 and 7 in terms of the annual and diurnal cycles of evolution of T_{mrt} and its corresponding thermal comfort condition in terms of PET index.

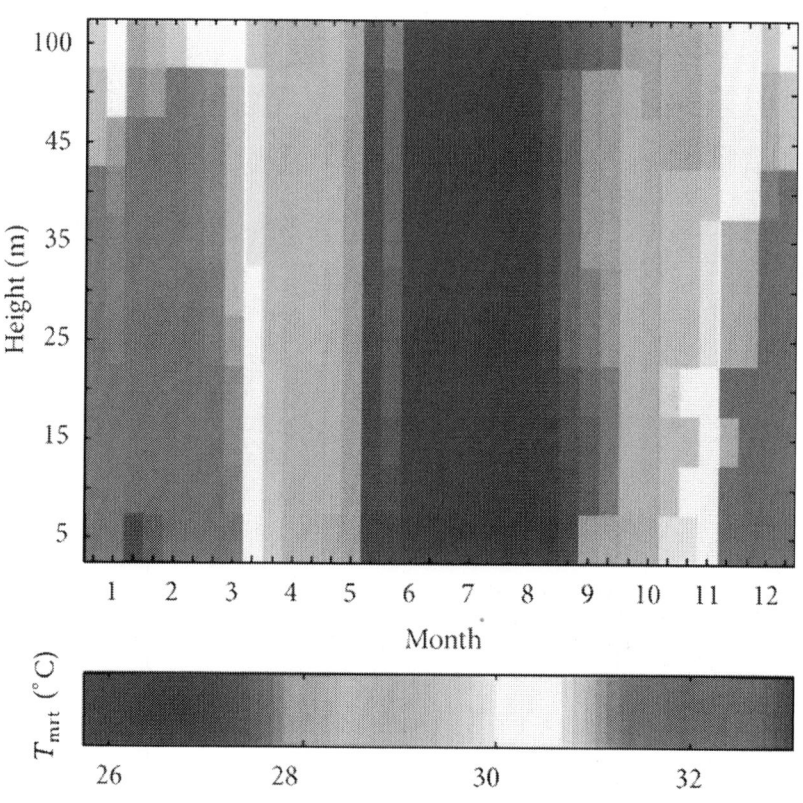

(a)

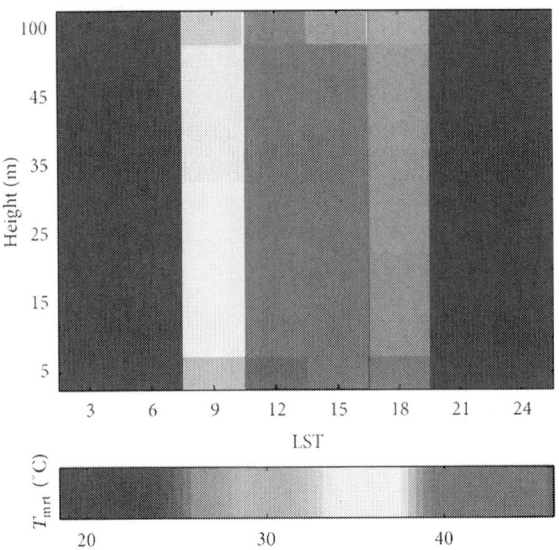

(b)

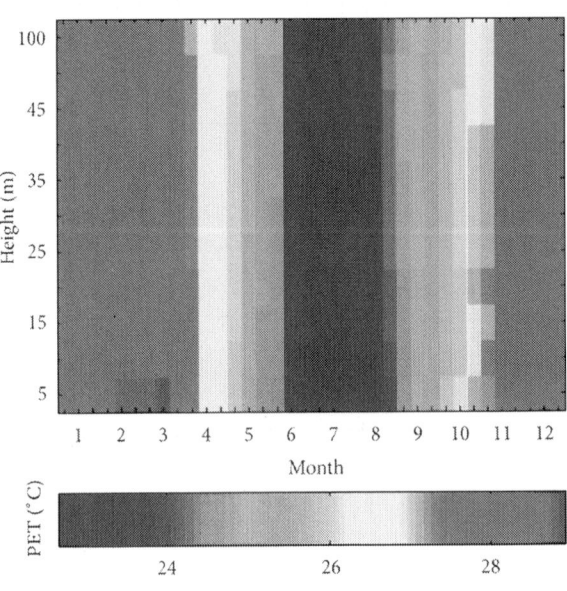

(c)

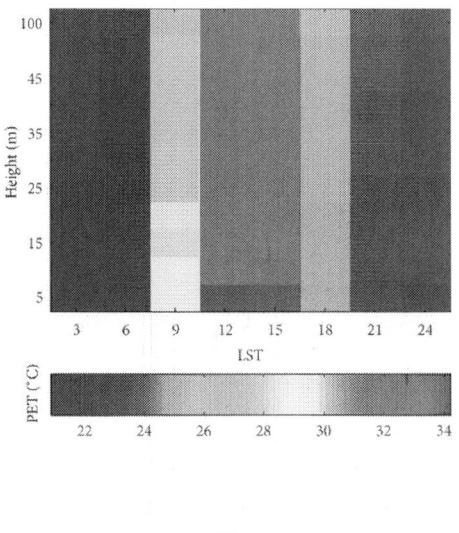

(d)

Figure 6: Effects of increasing buildings' height on the temporal distribution of T_{mrt} at the urban street as depicted in the annual cycle (a) and the diurnal cycle (b). The corresponding thermal comfort distribution is shown in terms of PET index ((c) and (d))

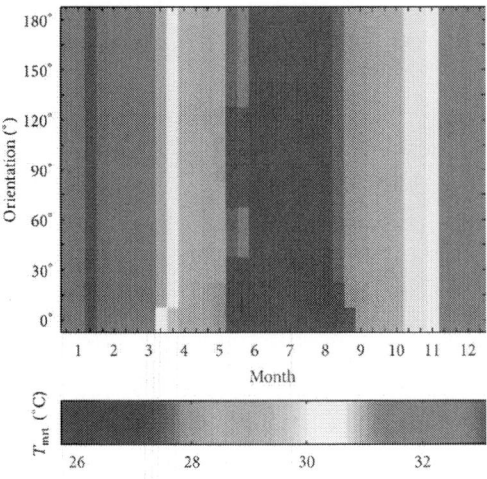

(a)

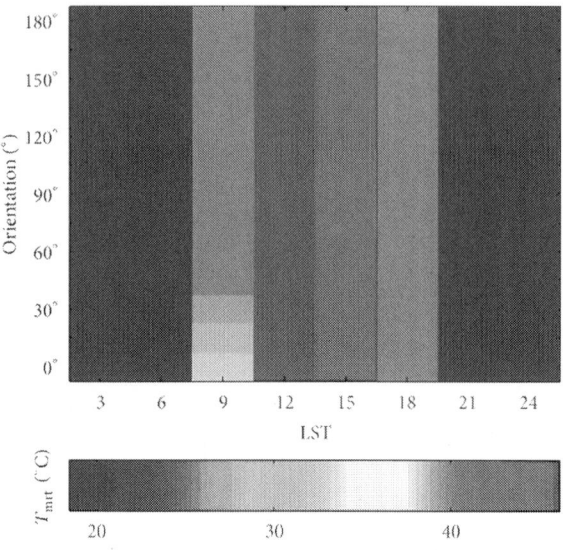

(b)

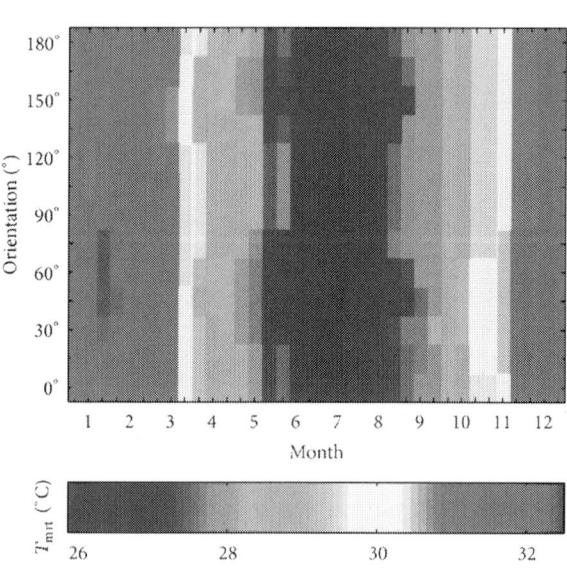

(c)

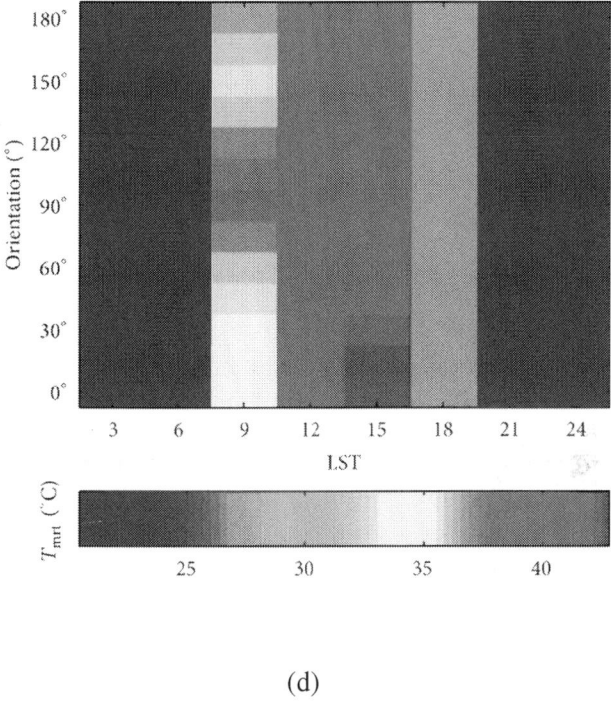

(d)

Figure 7: Effects of orientation on the temporal distribution of T_{mrt} at the urban street when the buildings' height was kept at 5 m, depicted in terms of the annual cycle of (a) and the diurnal cycle (b). (T_{mrt} (c) and (d)) show the corresponding effects when the buildings' height was kept at 25 m high.

From Figure 6, effect of buildings' height is described in terms of the annual and diurnal distribution of ((a) and (b)) and its corresponding PET index ((c) and (d)). It could then be evidently observed that a significant reduction of T_{mrt} and PET values just at the buildings' height of 20 m can be attained and more reduction at the height of 100 m. Whereas, effect of street orientation is illustrated by keeping the buildings at a height of 5 m (Figures 7(a) and 7(b)) and at a height of 25 m. Although significant differences could not be easily discerned at the 5 m buildings' height (Figures 7(a) and 7(b)), it is evident that street orientation particularly on the east-west direction (i.e., 45° in the figure) influences the calculated thermal comfort parameters as it can be observed in Figures 7(c) and 7(d). For this case, in Dar es Salaam, the east-west oriented street could be an undesirable street orientation for pedestrian thermal comfort condition. Other studies have attributed such an orientation to be prone to solar access [3, 6]. Otherwise, the north-south oriented streets (i.e., 120–135° in the figure) could present a good scenario for pedestrian thermal comfort in Dar es Salaam based on both annual and diurnal temporal distributions of T_{mrt}.

Simulation Results in the Urban Garden

Perhaps one of the reasons of doing simulation at the urban park is the understanding that many of the urban parks in Tanzanian cities including Dar es Salaam are relatively small in size but are also very close to surrounding buildings. The close buildings to such parks could potentially impose some significant effects on the thermal comfort in a garden or park (see MP2 in Figure 1) especially when buildings' height is too high. Figure 8, therefore, describes the effects of buildings' height on a small urban garden while the effect of orientation is illustrated in Figure 9.

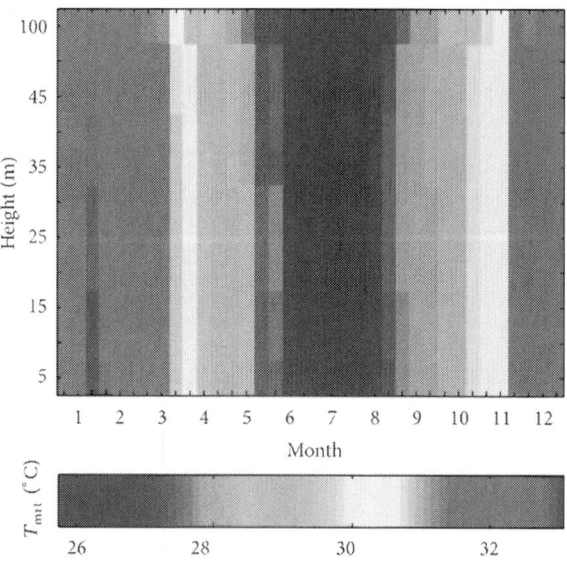

(a)

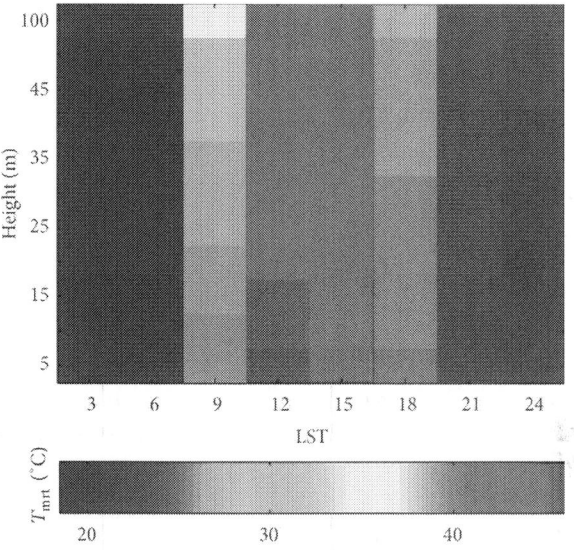

(b)

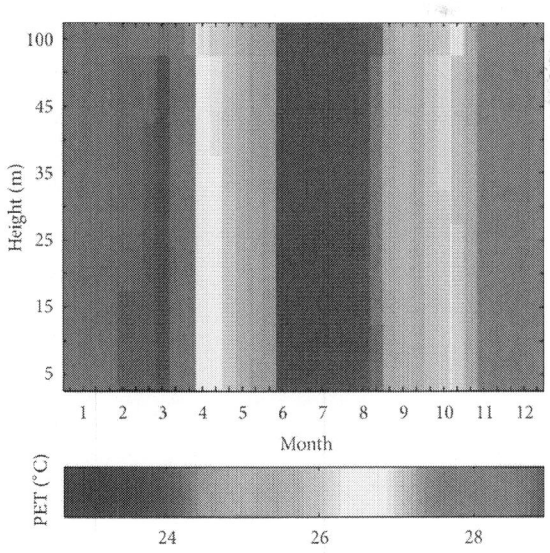

(c)

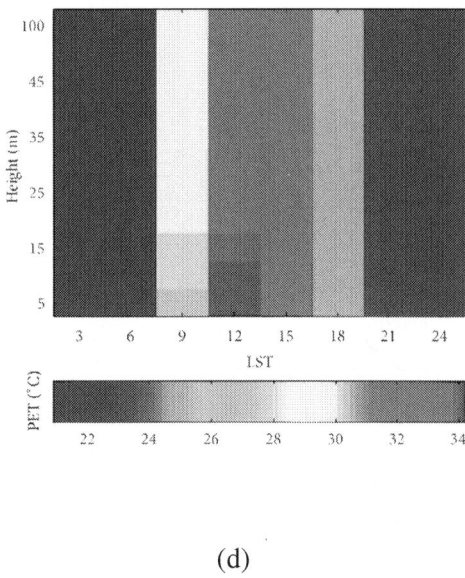

(d)

Figure 8: Effects of increasing buildings' height on the temporal distribution of T_{mrt} at the urban park as depicted in terms of the annual cycle (a) and diurnal cycle (b). The corresponding thermal comfort distribution is shown in terms of PET index ((c) and (d)).

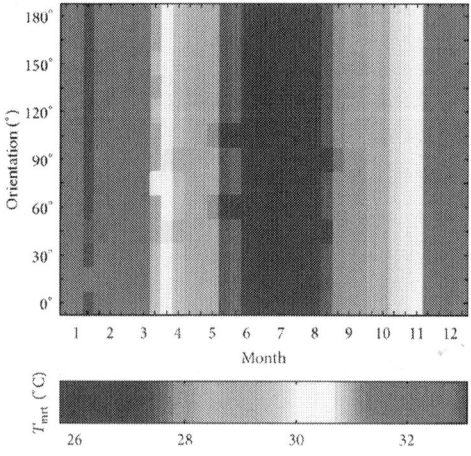

(a)

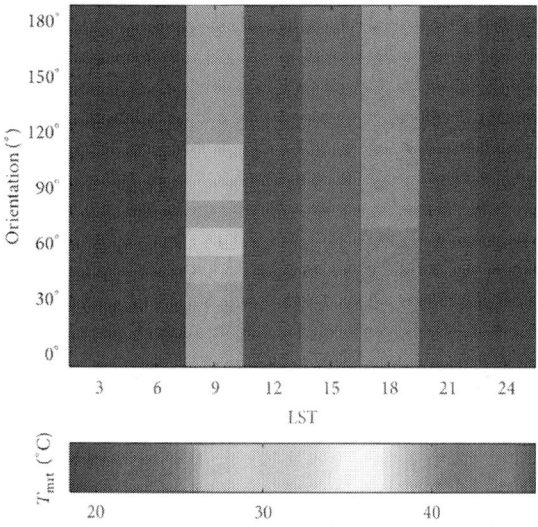

(b)

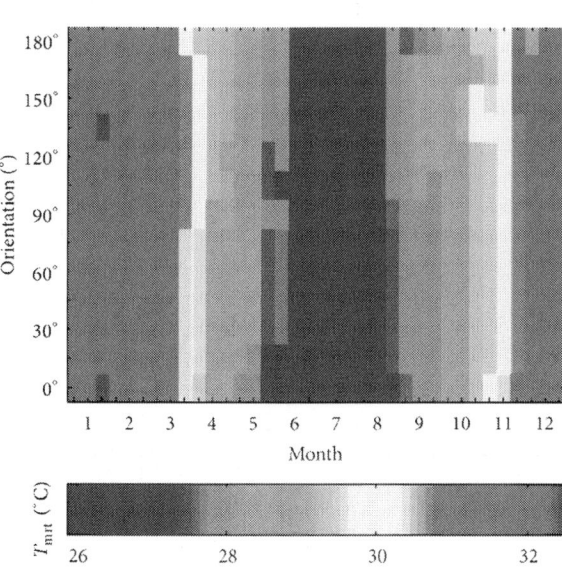

(c)

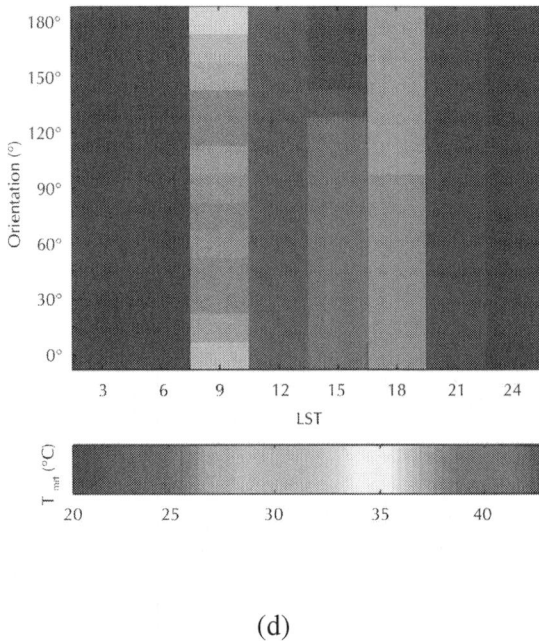

(d)

Figure 9: Effects of orientation on the temporal distribution of T_{mrt} at the urban park in terms of annual and diurnal cycles ((a) and (b)) with buildings' height kept at 5 m. Corresponding effect when buildings' height was increased to 100 m high ((c) and (d)).

The simulations results at the urban park suggested that buildings close to the park could significantly affect the thermal comfort at pedestrian level (i.e., human beings resting in an urban park), and the effects become clear as the height of the buildings increases. For instance, at a building height of 100 m, both T_{mrt} and PET values were observed to be significantly reduced (Figure 8). In order to visualize the buildings effect on the park thermal comfort, the park setup was also idealized and rotated every 15°. It should be noted that the buildings surrounding the urban parks are not usually uniformly located around the park as the case here. Results in this simulation are therefore illustrated in Figure 9.

In Figures 9(a) and 9(b), the buildings surrounding the park were kept at 5 m high, and it is clear that a significant reduction of T_{mrt} values could be reached when most of the buildings are located to the south-western side of the park (60° and 105° in the figure). In the diurnal distribution, this reduction could be revealed clearly at 9 LST. When the buildings' height was increased to 100 m, still the situation where most of them are located to the southwest side favours the significant reduction of T_{mrt} (Figures 9(a) and 9(b)). Due to the increased height of the buildings, the effect can also be observed at 15 LST on the diurnal cycle.

The simulations of varying the heights of surrounding buildings and rotating the urban park in order to determine the effect on thermal comfort at pedestrian level seems to be of a novelty in urban climate simulation studies. Many studies have concentrated on simulation of an urban canyon [3, 6, 11, 15, 38], and most of the studies suggested that at pedestrian level, the extreme cases in terms of thermal comfort parameters could be observed at the east-west and north-south orientations [3, 6, 38]. This situation was also observed in this study although with some slight differences which could be attributed to the latitudinal location of the cities which slightly affects the solar path of a particular place.

Simulation of thermal comfort for open spaces is very important especially in realizing the thermal comfort and recreation potential for the parks. Information like this has potential impact on design implications in cities particularly in developing countries where open spaces in cities might be considerably few. It can also help in management of urban parks in the sense that optimal comfort conditions within the urban park could be attained through shading effects of plants within the park itself and the influence of nearby buildings. In this study, buildings' height was particularly observed to exert a decisive role in altering the thermal comfort parameter values. Although, generally the increase in buildings' height was observed to favour reduction in thermal comfort parameters mainly through altering the solar access, this could also have an effect on the wind flow as already discussed in [15]. For Dar es Salaam example, any design affecting the city open spaces should consider taking the advantage of the sea breezes too.

CONCLUSIONS

Due to its economic importance in eastern and central Africa, Dar es Salaam city is growing fast. There are increasingly many tall buildings especially in the central business district. Although there are no continuous measurements of urban climate, simulation of urban climate especially to quantify the effects of building and street orientation at pedestrian level using the easily available synoptic data is important and could provide applicable information in terms of urban planning in developing countries. Simulation of urban climate at pedestrian level is also of crucial impact in tropical cities since most of the activities are outdoors. For instance, in Dar es Salaam, there are lot of street vendors who spend most of their daytime hours outside in streets. Thus, the results provided in this study could help in redesigning the streets that ensure thermal comfort at pedestrian level through the variation of buildings' height and orientation.

The simulations performed in this study shows that the thermal comfort parameters (T_{mrt} and PET) at both the urban street and the park can be significantly affected by the urban configuration. Optimal reduction of T_{mrt} and PET values could particularly be obtained on the north-south reoriented streets and with increased buildings' heights. Additionally, the results from the simulations undertaken on a small urban park provided a novelty in design implications and management of open spaces in many cities in the developing countries including Dar es Salaam. These should also help planners understand the potential of open spaces on improving the microclimates in urban areas. The results of this study could also ultimately assist planners in determining appropriate configuration in urban growth areas of Tanzania and elsewhere. Further urban spatial modelling of open spaces in Dar es Salaam is also suggested involving more microscale models for the sake of having robust and firm bioclimatic information to assist in urban planning and management of open spaces.

ACKNOWLEDGMENTS

The meteorological data used in the study were obtained from the Tanzania Meteorological Agency and fromhttp://www.ogimet.com/.

Emmanuel L. Ndetto would like to acknowledge the financial support granted from the joint scholarship program of Tanzania and Germany through the Germany Academic Exchange Service and Ministry of Education and Vocational Training (DAAD-MOEVT) for his stay in Freiburg, Germany.

REFERENCES

1.   S. Nieuwolt, "Breezes along the Tanzanian East Coast," Archiv für Meteorologie Geophysik und Bioklimatologie B, vol. 21, no. 2-3, pp. 189–206, 1973. · ·

2.   E. L. Ndetto and A. Matzarakis, "Basic analysis of climate and urban bioclimate of Dar es Salaam, Tanzania," Theoretical and Applied Climatology, 2013. ·

3.   F. Ali-Toudert and H. Mayer, "Thermal comfort in an east-west oriented street canyon in Freiburg (Germany) under hot summer conditions," Theoretical and Applied Climatology, vol. 87, no. 1–4, pp. 223–237, 2007. · ·

4.   I. Charalampopoulos, I. Tsiros, A. Chronopoulou-Sereli, and A. Matzarakis, "Analysis of thermal bioclimate in various urban configurations in Athens, Greece," Urban Ecosystems, vol. 16, no. 2, pp. 217–233, 2013.

5.   F. Bourbia and F. Boucheriba, "Impact of street design on urban microclimate for semi arid climate (Constantine)," Renewable Energy, vol. 35, no. 2, pp. 343–347, 2010. · ·

6.   F. Ali-Toudert and H. Mayer, "Numerical study on the effects of aspect ratio and orientation of an urban street canyon on outdoor thermal comfort in hot and dry climate," Building and Environment, vol. 41, no. 2, pp. 94–108, 2006. · ·

7.   T.-P. Lin, A. Matzarakis, and R.-L. Hwang, "Shading effect on long-term outdoor thermal comfort,"Building and Environment, vol. 45, no. 1, pp. 213–221, 2010. · ·

8.   E. Johansson, "Influence of urban geometry on outdoor thermal comfort in a hot dry climate: a study in Fez, Morocco," Building and Environment, vol. 41, no. 10, pp. 1326–1338, 2006. · ·

9.   H. Sugawara, A. Hagishima, K. Narita, H. Ogawa, and M. Yamano, "Temperature and wind distribution in an EW-oriented

urban street canyon," Scientific Online Letters of the Atmosphere (SOLA), vol. 4, pp. 53–56, 2008.

10. R. Emmanuel and E. Johansson, "Influence of urban morphology and sea breeze on hot humid microclimate: the case of Colombo, Sri Lanka," Climate Research, vol. 30, no. 3, pp. 189–200, 2006.

11. E. Johansson and R. Emmanuel, "The influence of urban design on outdoor thermal comfort in the hot, humid city of Colombo, Sri Lanka," International Journal of Biometeorology, vol. 51, no. 2, pp. 119–133, 2006. · ·

12. A. Lopes, S. Lopes, A. Matzarakis, and M. J. Alcoforado, "The influence of the summer sea breeze on thermal comfort in funchal (Madeira). A contribution to tourism and urban planning," Meteorologische Zeitschrift, vol. 20, no. 5, pp. 553–564, 2011. · ·

13. S. Ahmad, N. M. Hashim, Y. M. Jani, and N. Ali, "The impacts of sea breeze on urban thermal environment in tropical coastal area," Advances in Natural and Applied Sciences, vol. 6, no. 1, pp. 71–78, 2012.

14. E. L. Krüger and F. A. Rossi, "Effect of personal and microclimatic variables on observed thermal sensation from a field study in southern Brazil," Building and Environment, vol. 46, no. 3, pp. 690–697, 2011. · ·

15. R. Emmanuel, H. Rosenlund, and E. Johansson, "Urban shading—a design option for the tropics? A study in Colombo, Sri Lanka," International Journal of Climatology, vol. 27, no. 14, pp. 1995–2004, 2007. · ·

16. A. Matzarakis, F. Rutz, and H. Mayer, "Modelling radiation fluxes in simple and complex environments—application of the RayMan model," International Journal of Biometeorology, vol. 51, no. 4, pp. 323–334, 2007. · ·

17. A. Matzarakis, F. Rutz, and H. Mayer, "Modelling radiation fluxes in simple and complex environments: basics of the RayMan model," International Journal of Biometeorology, vol. 54, no. 2, pp. 131–139, 2010. · ·

18. S. Thorsson, F. Lindberg, I. Eliasson, and B. Holmer, "Different methods for estimating the mean radiant temperature in an outdoor urban setting," International Journal of Climatology, vol.

27, no. 14, pp. 1983–1993, 2007. · ·

19. P. Jonsson, C. Bennet, I. Eliasson, and E. Selin Lindgren, "Suspended particulate matter and its relations to the urban climate in Dar es Salaam, Tanzania," Atmospheric Environment, vol. 38, no. 25, pp. 4175–4181, 2004. · ·

20. S. E. Mbuligwe and G. R. Kassenga, "Automobile air pollution in Dar es Salaam City, Tanzania," Science of the Total Environment, vol. 199, no. 3, pp. 227–235, 1997. · ·

21. A. Hill and C. Lindner, Modelling Informal Urban Growth Under Rapid Urbanisation: A CA-Based Land-Use Simulation Model for the City of Dar Es Salaam, Tanzania, Technische Universität Dortmund, Dortmund, Germany, 2010.

22. F. R. Fosberg, B. J. Garnier, and A. W. Küchler, "Delimitation of the humid tropics," Geographical Review, vol. 51, no. 3, pp. 333–347, 1961. ·

23. M. Roth, "Review of urban climate research in (sub)tropical regions," International Journal of Climatology, vol. 27, no. 14, pp. 1859–1873, 2007. · ·

24. C. Howorth, I. Convery, and P. O›Keefe, "Gardening to reduce hazard: urban agriculture in Tanzania,"Land Degradation and Development, vol. 12, no. 3, pp. 285–291, 2001. · ·

25. W. Kuttler, "Stadtklimain," in Handbuch der Umweltveränderungen und Ökotoxologie, R. Guderian, Ed., pp. 420–470, Springer, Berlin, Germany.

26. A. Matzarakis, M. Rocco, and G. Najjar, "Thermal bioclimate in Strasbourg-the 2003 heat wave,"Theoretical and Applied Climatology, vol. 98, no. 3-4, pp. 209–220, 2009. · ·

27. M. G. Kendall, "A new measure of rank correlation," Biometrika, vol. 30, no. 1/2, pp. 81–93, 1938.

28. A. Matzarakis, H. Mayer, and M. G. Iziomon, "Applications of a universal thermal index: physiological equivalent temperature," International Journal of Biometeorology, vol. 43, no. 2, pp. 76–84, 1999.

29. H. Mayer and P. Höppe, "Thermal comfort of man in different urban environments," Theoretical and Applied Climatology, vol. 38, no. 1, pp. 43–49, 1987. · ·

30. P. Höppe, "The physiological equivalent temperature—a universal index for the biometeorological assessment of the

thermal environment," International Journal of Biometeorology, vol. 43, no. 2, pp. 71–75, 1999.

31. D. J. Wessel, Ashrae Fundamentals Handbook 2001, ASHRAE, Atlanta, Ga, USA, 2001.

32. A. Matzarakis, F. Rutz, and H. Mayer, "Modelling radiation fluxes in simple and complex environments—application of the RayMan model," International Journal of Biometeorology, vol. 51, no. 4, pp. 323–334, 2007. · ·

33. G. L. Makokha and C. A. Shisanya, "Trends in mean annual minimum and maximum near surface temperature in Nairobi City, Kenya," Advances in Meteorology, vol. 2010, Article ID 676041, 6 pages, 2010. ·

34. O. Akinbode, A. Eludoyin, and O. Fashae, "Temperature and relative humidity distributions in a medium-size administrative town in southwest Nigeria," Journal of Environmental Management, vol. 87, no. 1, pp. 95–105, 2008. · ·

35. R. Emmanuel, "Thermal comfort implications of urbanization in a warm-humid city: the Colombo Metropolitan Region (CMR), Sri Lanka," Building and Environment, vol. 40, no. 12, pp. 1591–1601, 2005. · ·

36. C. Grimmond, S. Potter, H. Zutter, and C. Souch, "Rapid methods to estimate sky-view factors applied to urban areas," International Journal of Climatology, vol. 21, no. 7, pp. 903–913, 2001. · ·

37. N. D. Burgess and G. P. Clarke, Coastal Forests of Eastern Africa, IUCN, Gland, Switzerland, 2000.

38. J. Herrmann and A. Matzarakis, "Mean radiant temperature in idealised urban canyons-examples from Freiburg, Germany," International Journal of Biometeorology, vol. 56, no. 1, pp. 199–203, 2012. · ·

Citations

CHAPTER 1

M. César, "Physics as Final Opportunity to Prevent Harms Related to Theatricalization of Meteorology," Atmospheric and Climate Sciences, Vol. 4 No. 1, 2014, pp. 131-136. doi: 10.4236/acs.2014.41015.

CHAPTER 2

Deen Mani Lal, Sachin D. Ghude, Jagvir Singh, and Suresh Tiwari, "Relationship between Size of Cloud Ice and Lightning in the Tropics," Advances in Meteorology, vol. 2014, Article ID 471864, 7 pages, 2014. doi:10.1155/2014/471864.

CHAPTER 3

Sharon E. Nicholson, "The West African Sahel: A Review of Recent Studies on the Rainfall Regime and Its Interannual Variability," ISRN Meteorology, vol. 2013, Article ID 453521, 32 pages, 2013. doi:10.1155/2013/453521.

CHAPTER 4

S. Tiwari, D. S. Bisht, A. K. Srivastava, G. P. Shivashankara, and R. Kumar, "Interannual and Intraseasonal Variability in Fine Mode Particles over Delhi: Influence of Meteorology," Advances in Meteorology, vol. 2013, Article ID 740453, 9 pages, 2013. doi:10.1155/2013/740453.

CHAPTER 5

Diofantos G. Hadjimitsis, Adrianos Retalis, Silas Michaelides, Filippos Tymvios, Dimitrios Paronis, Kyriacos Themistocleous and Athos Agapiou (2013). Satellite and Ground Measurements for Studying the Urban Heat Island Effect in Cyprus, Remote Sensing of Environment - Integrated Approaches, (Ed.), ISBN: 978-953-51-1152-8, InTech, DOI: 10.5772/39313.

CHAPTER 6

Eric A. Hendricks and Melinda S Peng (2012) Initialization of Tropical Cyclones in Numerical Prediction Systems, Advances in Hurricane Research - Modelling, Meteorology, Preparedness and Impacts, Dr. Kieran Hickey (Ed.), ISBN: 978-953-51-0867-2, InTech, DOI: 10.5772/51177.

CHAPTER 7

Emmanuel Lubango Ndetto and Andreas Matzarakis, "Effects of Urban Configuration on Human Thermal Conditions in a Typical Tropical African Coastal City," Advances in Meteorology, vol. 2013, Article ID 549096, 12 pages, 2013. doi:10.1155/2013/549096.

Index

A

African easterly jet (AEJ) 57, 73
African Easterly Jet (AEJ) 50, 72
African Easterly Waves (AEWs) 50
African Westerly Jet (AWJ) 72, 77
Air Force Weather Agency (AFWA) 209
Artificial Neural networks (ANN) 176

C

Central business district (CBD) 239
Central Pollution Control Board (CPCB) 142
Compressed natural gas (CNG) 149
Coriolis force 8

D

December–February (DJF) 222
Democratic Republican of Congo (DRC) 27
Dropwindsonde Observations for Typhoon Surveillance near the Taiwan Region (DOTSTAR) 210
Dynamic initialization (DI) 212

E

Earth Resources Observation and Science (EROS) 188
European Center for Medium Range Weather Forecasting (ECMWF) 206

G

Generalized split-window (GSW)
180
Generalized Split Window (GSW)
176
Geostrophic winds 10

H

Human body 228
Hurricane Forecast Improvement
Project (HFIP) 196

I

Income Tax Office (ITO) 142
Indo-Gangetic Plains (IGP) 142
Intertropical convergence zone
(ITCZ) 225
Intertropical Convergence Zone
(ITCZ) 57, 108

J

Julius Nyerere International Airport
(JNIA) 227

L

Land surface temperature (LST)
166, 179
Land Surface Temperature (LST) 176
Lightning Imaging Sensor (LIS) 22

M

Mean sea level pressure (MSLP)
204, 208
Mesoscale convection systems
(MCSs) 93
Mesoscale Convective Systems
(MCSs) 47
Meteorological Organization 170
Moderate Resolution Imaging Spec-
troradiometer (MODIS) 22
Multi-Layer Perceptron (MLP) 177

N

National Ambient Air Quality Moni-
toring Network (NAAQMN)
142
National Ambient Air Quality Stan-
dards (NAAQS) 140
National Centers for environmental
Prediction (NCEP) 206
National Oceanic and Atmospheric
Administration (NOAA) 196

O

Optimal interpolation (OI) 205

P

Physiologically equivalent (PET)
222
Physiologically equivalent temperature (PET) 224

R

Relative humidity (RH) 143, 150

S

Saharan Heat Low (SHL) 79
Science instrument 23
Surface urban heat island (SUHI)
166

T

Thermodynamic constraint 12
Three-dimensional variational
(3DVAR) 204
Total suspended particulate matters
(TSPM) 146
TRMM Microwave Instrument (TMI)
22
TRMM Online Visualization and
Analysis System (TOVAS) 22
Tropical cyclones (here after TCs)
195
Tropical easterly jet (TEJ) 57, 73

Tropical Easterly Jet (TEJ) 50, 72, 118

Tropical Rainfall Measurement Mission's (TRMM) 22

U

Urban heat island (UHI) 165, 167, 180

U.S. Geological Survey (USGS) 188

W

Weather Research and Forecasting (WRF) 209

West African Westerly Jet (WAWJ) 72, 78

Wind speed (WS) 143, 150